Advanced Wastewater Treatment

Advanced Wastewater Treatment

Russell L. Culp

General Manager
South Tahoe Public Utility District
South Lake Tahoe, California

and

Gordon L. Culp

Manager
Water and Waste Management Section
Battelle-Northwest
Richland, Washington

Van Nostrand Reinhold Company

New York/Cincinnati/Toronto/London/Melbourne

Van Nostrand Reinhold Company Regional Offices:
New York Cincinnati Chicago Millbrae Dallas
Van Nostrand Reinhold Company International Offices:
London Toronto Melbourne

Library of Congress Catalog Card Number: 78-147192

Manufactured in the United States of America
Published by Van Nostrand Reinhold Company
450 West 33rd Street, New York, N.Y. 10001
Published simultaneously in Canada by Van Nostrand Reinhold Ltd.
15 14 13 12 11 10 9 8 7 6 5 4 3 2 1

VAN NOSTRAND REINHOLD
ENVIRONMENTAL
ENGINEERING SERIES

ADVANCED WASTEWATER TREATMENT, by Russell L. Culp and Gordon L. Culp

ARCHITECTURAL INTERIOR SYSTEMS—Lighting, Air Conditioning, Acoustics, John E. Flynn and Arthur W. Segil

THERMAL INSULATION, by John F. Malloy

INDUSTRIAL WASTE DISPOSAL, edited by Richard D. Ross

MICROBIAL CONTAMINATION CONTROL FACILITIES, by Robert S. Runkle and G. Briggs Phillips

SOUND, NOISE, AND VIBRATION CONTROL, by Lyle F. Yerges

VAN NOSTRAND REINHOLD
ENVIRONMENTAL
ENGINEERING SERIES

THE VAN NOSTRAND REINHOLD ENVIRONMENTAL ENGINEERING SERIES is dedicated to the presentation of current and vital information relative to the engineering aspects of controlling man's physical environment. Systems and subsystems available to exercise control of both the indoor and outdoor environment continue to become more sophisticated and to involve a number of engineering disciplines. The aim of the series is to provide books which, though often concerned with the life cycle—design, installation, and operation and maintenance—of a specific system or subsystem, are complementary when viewed in their relationship to the total environment.

Books in the Van Nostrand Reinhold Environmental Engineering Series include ones concerned with the engineering of mechanical systems designed (1) to control the environment within structures, including those in which manufacturing processes are carried out, (2) to control the exterior environment through control of waste products expelled by inhabitants of structures and from manufacturing processes. The series will include books on heating, air conditioning and ventilation, control of air and water pollution, control of the acoustic environment, sanitary engineering and waste disposal, illumination, and piping systems for transporting media of all kinds.

Preface

The control of water pollution in the United States and in many other countries undoubtedly will require treatment techniques far more efficient and reliable than the conventional processes considered as "complete" treatment in the past. There are many instances today where removal is needed of materials such as phosphorus or refractory organics, which are not removed by conventional processes, in order to alleviate pollution or to permit reuse of treated wastewaters. There will be many more such instances in the near future. The techniques are available now to renovate wastewaters to any degree desired, at reasonable costs in most cases. Certainly, it is hoped that future development will reduce costs. In many cases, there is not time to wait for this hoped-for result. Unfortunately, there is a substantial gap between available, proven technology and that which is being expeditiously brought to bear by the technical and political forces seeking to control pollution. It is the purpose of this book to present the basic principles, engineering design information, and actual operating experiences related to treatment techniques which are relatively new to the wastewater treatment field, with the hope that it will assist in closing the existing gap in practical application of new technology. The treatment methods discussed are restricted to those designed to remove pollutants normally remaining after conventional secondary treatment.

The authors of this book have been fortunate to be involved with a most advanced wastewater renovation project since its inception in the early 1960's. This project is located at South Lake Tahoe, California, where a new plant converts municipal wastewater into water pure enough to support a rainbow trout fishery and to permit unrestricted recreational activities in a reservoir made up solely of the treated wastewater. The experimental work, design work, and subsequent plant-scale experiences at South Lake Tahoe are referred to frequently in this book. The work of other investigators at other locations is also drawn on heavily to present, as nearly as possible, an accurate review of currently available treatment methods. It is hoped that the resulting compilation will be of value to design engineers as well as to students and pollution control authorities, and that it will hasten the more widespread application of new methods for processing wastewater.

In preparing any technical book, there is the risk that some material may be outdated by the time that the book is in print. This risk is particularly hazardous in a field receiving as much research attention as the one described. However, the bulk of the material presented describes unit processes which are likely to be used for many years, although they may be combined with other unit processes not yet developed. The major portion of this book will deal with processes that have been demonstrated successfully on a plant scale. A brief discussion of promising processes not yet beyond the pilot plant or laboratory scale is also included. In order to prepare a volume of reasonable size, the treatment of some subjects may be briefer than the reader desires. Appropriate references are included so that the reader may pursue these subjects in greater depth if he chooses to do so.

Special thanks are expressed to the consulting engineering firms of Clair A. Hill & Associates and Cornell, Howland, Hayes & Merryfield, the design engineers for the South Lake Tahoe project, for their permission to cite specific design details from their work. Also, without the unusual vision and determination of the Board of Directors of the South Tahoe Public Utility District many plant-scale experiences related in this book would not have been available. Several equipment manufacturers were generous in supplying illustrative material for the text. The encouragement offered by our families and colleagues throughout the preparation of this book was essential to its completion.

RUSSELL L. CULP
GORDON L. CULP

January, 1971

Contents

1

Purpose and Benefits of Advanced Wastewater Treatment

PURPOSE

Advanced wastewater treatment technology is designed to remove pollutants which are not adequately removed by conventional secondary treatment processes, previously considered "complete" processes. These pollutants may include soluble inorganic compounds such as phosphorus or nitrogen, which may support algal growths in receiving waters; organic materials contributing biochemical oxygen demand, chemical oxygen demand, color, taste, and odor; bacteria; viruses; colloidal solids contributing turbidity; or soluble minerals which may interfere with subsequent reuse of the wastewater. The purpose of advanced waste treatment may be to alleviate pollution of a receiving watercourse or to provide a water quality adequate for reuse, or both. The advanced waste treatment process may be used following, in conjunction with, or replace entirely the conventional secondary process.

Increasing population and increasing water use has already created, in many locations, pollution problems which cannot be adequately solved by secondary treatment. It is inevitable that the number of these instances will increase in the future. It is also inevitable that the deliberate reuse of treated wastewaters will be required to meet future water demands. Indirect water reuse is already commonly practiced, with some estimates indicating that 40 percent of the United States population is using water that has been used at least once before for domestic or industrial purposes. This indirect reuse will also increase in the future. All of these factors indicate that use of advanced wastewater treatment techniques will become increasingly common.

PUBLIC ATTITUDE AND NATIONAL SIGNIFICANCE

The public attitude toward pollution control, which bordered on apathy during the first half of the twentieth century, will undergo drastic change in the early 1970's based on the recent surge in public concern for the quality of the environment. The strong desire of the people for adequate pollution control programs is reflected in the overwhelming margins by which very large pollution control bond issues have recently carried. For example, the voters of the state of New York approved a $1 billion bond issue for pollution control in 1968 by a margin of 4–1. The people of St. Louis approved a $95 million pollution bond issue by a 5–1 margin. These results reflect the intense public desire to improve our environment. Sincere and concerted public concern will be required over a long period of time to make the necessary changes in society to bring about significant improvements in our environment. Much more than clever technological advances will be needed. Major changes in our political, social, legal, and economic approaches to pollution control will be required.

All municipal wastewater could be completely eliminated as a source of pollution in the United States and converted to a quality adequate to provide a valuable water resource for nearly unrestricted reuse at a national annual cost of only about 75 cents per person per month.

Proper application of existing technology could reduce this cost by 30 percent through construction of physical-chemical treatment plants providing advanced waste treatment of raw sewage rather than construction of additional secondary treatment plant capacity. It is not suggested that such nonselective use of advanced wastewater techniques is the optimum economic solution to achieving environmental goals but this example illustrates what could be accomplished with available resources by the establishment of appropriate national goals.

CHARACTERISTICS OF SECONDARY EFFLUENTS

The current federally established goal of universal secondary treatment will result in temporary improvement in water quality in some areas while merely holding the line in others. It is only a question of time before conventional secondary processes operating at their highest efficiency will be inadequate in a significant portion of the United States. Materials present in effluent from a well-operating secondary plant which may be of concern can be placed in the general categories of soluble organic compounds, soluble inorganic compounds, particulate solid material, and pathogenic organisms.

Organic Compounds

Efficient secondary processes employ biological treatment to remove essentially all of the soluble, biologically degradable organic material in municipal wastewaters. A portion of the soluble organics removed are converted to biological organic cell material which in turn, can exert oxygen demand in the effluent. Generally, the net removal of bio-degradable organics is on the order of 90 percent. The remaining degradable organics will exert a demand on the oxygen resources of the receiving body of water which may or may not have an adverse effect, depending on the assimilative capacity of that body of water.

Nondegradable organics are, of course, not removed by secondary processes using bio-degradation techniques. These organics can cause taste and odor problems in downstream water supplies. They also impart a color to the effluent which may make it unsuitable for many direct-reuse applications and may make the receiving stream aesthetically unacceptable for recreation. In some cases, they may cause objectionable tastes in fish residing in the receiving watercourse. They may also pass through downstream water treatment plants, causing yet-unknown long-term physiological effects on downstream water users. In some cases, they may cause foam in a receiving stream although the introduction of bio-degradable detergents has done much to reduce this problem.

Inorganic Compounds

Phosphorus and nitrogen are two key elements required by algae for growth which are not significantly removed by conventional secondary processes. Phosphates also may interfere with the coagulation processes used in downstream water treatment plants. A major source of phosphorus is the phosphate-builders used in modern detergents. The growth of algae in a receiving body of water may create aesthetically unacceptable conditions for recreation, may create taste and odor problems in downstream water supplies, may cause operating problems in downstream water filtration plants, and may create a significant oxygen demand during night time hours or after death of the algae. There remains some debate on the minimum phosphorus and nitrogen concentrations which will support objectionable algal growths. However, removal of phosphorus to a concentration of about 0.1 mg/l has reduced algal growths to an insignificant level in a reservoir made up solely of effluent from the South Lake Tahoe wastewater reclamation plant.

During use of water in a municipality, the mineral quality of the water is altered. Inorganic salts containing calcium, magnesium, sodium, potassium, chlorides, sulfates, and phosphates are among those added. Normal water treatment practices at downstream locations do not remove these salts. As a result, the dissolved solids content increases as a fixed supply source passes through several users in series. It is generally agreed that 500 mg/l of dissolved solids is the upper limit for palatable water. Excessive dissolved solids concentrations can cause laxative action in the user, although no harmful permanent physiological effects are known. Dissolved solids concentrations can also adversely affect irrigation use, industrial use, or stock and wildlife watering. Calcium and magnesium contribute to downstream water hardness.

Particulate Solids

Although an efficient secondary plant removes 90–95 percent of the incoming suspended solids, much poorer removals occur all too frequently during "upsets" of the secondary plant due to poor operation, hydraulic or organic overloads, or mechanical failures. The 90–95 percent removal is not adequate for many direct reuse applications. Suspended solids can interfere with disinfection of the effluent, leading to the discharge of pathogenic organisms. In cases of gross secondary plant failures and small receiving streams, sludge deposits may result which can exert long-term oxygen demands in addition to being aesthetically unacceptable. The historically inconsistent performance of secondary plants is a major weakness which can be overcome with proper application of advanced waste treatment techniques to remove all suspended solids.

Pathogenic Organisms

Secondary processes provide substantial reductions in incoming viral and bacterial concentrations, but it has been shown that both viruses and bacteria are normally present in secondary effluents. Infectious hepatitis has been confirmed to be a waterborne viral disease with the virus capable of surviving 10 weeks in clean water. Certainly, direct reuse of secondary effluent for many purposes is not acceptable due to the health hazard.

UNAVOIDABLE WATER REUSE

The location of several cities on a single river, stream, or lake leads to unavoidable water reuse when each city uses the water body as a water supply and receiving body for wastewater. A recent study by the American Water Works Association showed that, on the average, one gallon out of every thirty used for water supply had passed through the wastewater system of an upstream community for the 155 cities studied (Haney, 1969). Streams containing water used by many upstream industries, agriculturists, and communities serve as sources of recreation and water supply. Unavoidable water reuse is an already well-established fact and accepted practice, although not often recognized as being so.

INTENTIONAL WATER REUSE

The need for direct and deliberate reuse of reclaimed sewage effluents is increasing in many areas of the world. Reuse holds the key to the efficient and effective utilization of the limited fresh water resource by making available a new valuable source of water to augment existing supplies and a major source of supply for the future. It is generally recognized that we are not running out of water, but have reached the point where maximum utilization of the available supply through elimination of wasteful and reckless degradation of the resource quality must occur if we are to provide for the water needs of the future.

The time is rapidly approaching in the United States when the degree of wastewater treatment required in many areas for pollution control will be so costly that cities will not be able to afford the luxury of discarding once-used water. In many other areas of the world, it is only a matter of time before serious consideration must be given to direct reclamation and reuse of sewage effluent to supplement inadequate potable water supplies.

To illustrate the expected need for extensive water reuse, by 1980 the available daily supply in the United States will be 515 billion gallons, if careful resource management is practiced and increased impoundment capacity is provided, but the demand will have increased to 650 billion gallons. To meet the total demand, it is apparent that there will have to be a considerable amount of reclamation and reuse of "secondhand" waters. At the present time it is estimated that the total of used municipal water is approaching 20 billion gallons daily (bgd) and that over 40 percent of this rapidly growing supply could be readily reclaimed for industrial use. There is the potential that water quality suitable for direct potable use can be produced from sewage effluents when there is sufficient economic justification. For example, complete utilization of available potable water supplies has forced the city of Windhoek, South Africa, to turn to direct potable use of reclaimed waste water (Cillie, *et al.*, 1966).

The increasing awareness of the value of reclaimed water as a potential resource is reflected by the following statement of policy made by the Water Pollution Control Federation in 1963: "Wastewater represents an increasing fraction of the nation's total water resource and is of such value that it might well be reclaimed for beneficial reuse through the restoration of an appropriate degree of quality. To this end, the development of methods for wastewater reclamation and criteria for such reuse should be encouraged." However, in spite of this apparent need for reuse, the process of recovering fresh water from the ocean continues to receive the greater public interest. Claims that the ocean can afford an inexhaustible supply of fresh water at a reasonable price are not likely to be realized, although desalting may provide an important supplemental supply for communities situated on the seacoast. In contrast to municipal wastewater, which contains less than 0.1 percent of impurities that can be effectively removed with existing advanced waste treatment methods, seawater contains 3.5 percent of dissolved salts plus considerable organic matter—over thirty-five times as much foreign matter as secondary sewage effluent. For these reasons capital investment and unit costs for desalination of sea water exceed costs for wastewater reclamation. The benefits to be gained from advanced waste treatment and water reuse include not only supplementation of fresh water supplies, but also alleviating, in the same step, the pollution problem. Reclamation of wastewaters eliminates the source of pollution and qualifies as complete pollution control.

The importance of wastewater reclamation and reuse in the United States received detailed consideration as early as 1947. At this time, a symposium on Reclamation of Sewage Effluents was held at a joint meeting between the Federation of Sewage Works Associations and the American Water Works Association. It was concluded that the economics of water reuse hinged on the value of the reclaimed water itself, rather than on the value of any impurities extracted from sewage. Some concern about the need for greater water reuse was expressed through the 1950's, but as the decade came to a close, the lack of severe water shortages in the United States provided little stimulation to take concrete action to encourage extensive reuse.

In 1947, sewage effluents were being utilized at 135 locations in the United States, of which 124 were for irrigation; 10 for industrial cooling, boiler feedwater, quenching and process water; and in one instance for a skating pond. The volume of water reused amounted to 0.3 percent of the daily sewage flow in the United States. As of 1954, it was estimated that utilization of sewage effluents in the United States did not exceed 1 percent of the total sewage flow. Reuse as irrigation water far exceeds that for other purposes.

Industrial Use

Industrial use of water in the United States has been estimated to be about seven times that of municipal use and about equal to the irrigation use. It has been estimated that by 1980, industrial water use will increase by 146 percent, municipal use by 68 percent, and irrigation by 18 percent over 1960 levels. It is apparent that as the demand for and cost of fresh water supplies increases, that wastewater reclamation will be an economically attractive alternate for industry in an increasing number of instances. The required quality of water varies widely depending upon the specific use involved. Required cooling water quality differs greatly from that for boiler feedwater, for example. Water quality exceeding drinking water standards is needed in some industrial applications such as certain electronic manufacturing operations, for high-pressure boiler feedwater, and in beverage preparation. It is impossible to tabulate the specific criteria required for all potential industrial uses. When considering a specific instance of industrial use of reclaimed wastewater, one should determine the needed water quality and evaluate the economics of obtaining this quality from a wastewater stream as compared to other alternate sources. Major items to be considered are the consistency in quality and quantity of each source.

Municipal wastewater treatment plant effluents have been used as industrial cooling water in several locations. The major considerations are that the water should not encourage corrosion, deposition of scale, delignification of wooden cooling towers, growth of microorganisms, and excessive foaming in cooling towers. Municipal wastewater temperatures usually show less variation in temperature than surface waters, although wastewaters frequently have higher temperatures.

Direct use of secondary effluents as cooling water has proven satisfactory in some cases, although chemical coagulation and sedimentation of secondary effluent is often provided prior to use as cooling water. If lime is used as a coagulant, care must be taken to insure that the *p*H is adjusted to a satisfactory value (6.5–7.5) to avoid delignification of wooden packings in cooling towers and to avoid deposition of calcium carbonate in the cooling system.

Wastewater constituents which may prove troublesome in cooling water supplies but which are readily controlled by advanced wastewater treatment processes are suspended solids, hardness agents, dissolved organics, dissolved oxygen or gases, algal nutrients, and slime-producing organisms. Turbidity and hardness values of less than 50 mg/l are generally desirable. Chemical additives to inhibit scale formation and slime growths may be used in conjunction with the wastewater treatment process.

Almost all industries use boilers and consequently have the problem of adequate boiler feedwater. The subject of required quality of boiler feedwater is a complex one which has been discussed in many published papers. As the operating pressure of the boiler increases, the quality of boiler feedwater must improve. For boilers of very high pressure (1,000 psi or more), all hardness must be removed, and dissolved solids should be as low as possible, preferably less than 0.5 mg/l. Carefully deionized and deaerated water is required in these cases. Silica is especially troublesome because it forms a hard scale in boilers and boiler tubes. For high-pressure boilers, dissolved silica cannot exceed 0.2 mg/l while 5 mg/l can be tolerated at pressures of 250–400 psi. No ammonia should be present because of damage to copper parts. High-pressure boilers require treatment steps beyond those normally used for advanced waste treatment purposes to remove soluble minerals of concern.

Bethlehem Steel Company has used Baltimore's wastewater for process purposes for over 20 years. Wolman (1948) reported the desirable water characteristics for steel manufacture as follows: temperature below 75°F, chlorides below 175 mg/l, *p*H between 6.8 and 7.0, hardness below 50 mg/l, suspended matter below 25 mg/l, organic content as low as possible, and corrosion potential as low as possible. The copper and aluminum industries also have several process steps in which reclaimed wastewater would be of suitable quality.

The phosphorus removal provided by the processes described in this book will reduce the algal growths which could interfere with oil well repressuring using stored effluent. Filtration is required to insure that suspended solids will not plug the injection well.

Industrial use of renovated municipal effluent for processing of foods or ice for human consumption will be limited by the same concerns discussed in a subsequent section of this chapter on domestic reuse. Fresh water sources will remain preferable for the foreseeable future for these purposes. There are other innumerable industrial applications where the processes described in this book are capable of providing adequate water quality.

Agricultural Use

Irrigation use is second only to industrial use of water in the United States and comprised about 43 percent of all water use in 1960 (compared to 50 percent by industry). As noted earlier, use of effluent for irrigation has been the major instance of wastewater reuse in the past in the United States. The chief concern has been to restrict the use for irrigation to prevent health hazards from effluent contact with crops directly consumed by humans. The use of advanced wastewater treatment techniques offers the potential for unrestricted, direct reuse for irrigation. Unrestricted reuse following groundwater recharge is an already-accepted practice. Recent studies by Bouwer (1968) have shown that secondary effluent from Phoenix, Arizona, can be renovated adequately to permit unrestricted irrigation by groundwater recharge through the normally dry Salt River bed.

The salt content of the irrigation water is of concern as well as the sanitary quality, but absolute limits of desirable salt content cannot be fixed because plants vary widely in their tolerance of salinity and soil types, climatic conditions, and irrigation practices. The relationship between ions may be significant, as illustrated by the antagonistic influence between calcium and sodium. The concentration of salts in most municipal effluents will not usually be high enough to cause immediate injury to crops. If leaching of the root zone does not take place, the salt concentration will increase until it reaches the limit of solubility of each salt. Water containing up to 2,000 mg/l of dissolved solids is suitable for most plants except sensitive ones. Dissolved solids of less than 1,000 mg/l are suitable for all types of plants provided drainage is good.

Calcium and magnesium in proper proportion maintain soil in good condition, while the opposite is true when sodium predominates. When the percentage of sodium (calculated as $\frac{Na \times 100}{Na + Ca + Mg + K}$ as milliequivalents [meq] per liter) exceeds the desirable limit for a given soil, granular soil structures begin to break down when the soil is moistened. The soil pores eventually seal, resulting in a decrease in soil permeability. It has been widely recommended that the percentage of sodium in irrigation water should not exceed 50–60 percent to avoid these deleterious effects. Where the soil has a high cation exchange capacity, higher percentages may be acceptable. The sodium hazard is increased if the water contains a high concentration of bicarbonate ions because there is a tendency for calcium and magnesium to precipitate as carbonates and for the relative proportion of sodium to be increased. It has been suggested that waters containing less than 1.25 meq/l of residual sodium carbonate (calculated as $[CO_3 + HCO_3] - [Ca + Mg]$) are probably safe, while those containing 1.25–2.5 meq/l are marginal (*Water Quality Criteria,* McKee, 1963).

Certain soluble salts can be harmful if present in excessive quantities. For example, 700–1500 mg/l of chlorides in the root zone can cause leaf burns and possibly death of the plant. Although most municipal wastewaters will be of adequate chemical quality for irrigation, careful evaluation of this aspect as well as of the potential health hazards is required. One should also be alert for toxic compounds originating from industrial wastes which may be discharged to the municipal sewer system.

Irrigation of golf courses and parks with secondary effluent has long been practiced in the Southwest. More advanced treatment may be preferable in many cases to minimize health hazards due to chance contact or by wind spray striking drinking fountains. Piping carrying the effluent must be clearly marked to prevent cross-connection or potable use. The need for careful control of these irrigated areas is apparent if direct contact with the effluent is to be prevented during irrigation periods.

Where soil and hydrologic conditions are proper, groundwater recharge through surface spreading of secondary effluent offers an economical method of providing an effluent for unrestricted irrigation. As noted earlier, a project of this type has been found adequate for unrestricted irrigation reuse at Phoenix, Arizona.

The practical significance of water used for stock watering as a vector of animal disease has not yet been established. Until more data are available, complete disinfection should be provided for water used by cattle. There is the risk that udders of cows having access to contaminated effluent or effluent-irrigated fields may become infected by typhoid or tuberculosis bacteria if disinfection of the effluent is not complete.

Domestic Use

There are some significant differences of opinion in different parts of the world on the suitability of direct reuse of wastewaters for a potable water supply. The city of Windhoek, South Africa, is successfully recycling advanced wastewater treatment plant effluent directly to the inlet of a water treatment plant. However, it is the present position of the U.S. Public Health Service that renovated sewage should not in any case be used as a source of drinking water when other sources are available. There are a number of good reasons to delay the direct reuse of renovated wastewaters for potable supplies until it is absolutely necessary. The Public Health Service Drinking Water Standards are based on the assumption that a sanitary survey has shown the raw water source is relatively unpolluted. These standards appear to be reliable for supplies taken from fresh water sources since acute health effects have not yet been attributed to waters that have met these standards. However, these standards may not be adequate for reclaimed wastewaters due to the presence of trace organic materials such as pesticides, antibiotics, hormones, or trace materials from industrial wastes. Also, the potential for viral contamination has not been accurately determined over a long period for various advanced wastewater treatment unit process combinations.

An instance of drought-forced recycle of trickling filter effluent through a 17-day retention pond directly to a softening plant intake in Chanute, Kansas, has received widespread attention (Meltzer, *et al.*, 1958). Although no health problems have been traced to this direct reuse, the presence of soluble materials not removed by secondary treatment led to poor consumer acceptance due to foaming and a pale yellow color in the tap water. Had advanced waste treatment facilities been practical, there is little doubt that the residents of Chanute would have been willing to pay many times the cost of normal water and sewage treatment to obtain a satisfactory water from their wastewater. Despite the lack of health problems at Chanute with rather crude techniques and at Windhoek with advanced technology, it appears that the most prudent approach at the present is to use reclaimed wastewaters for the many nonpotable uses, where it is unquestionably of adequate quality, to increase the availability of fresh water supplies for potable use. Indirect reuse through groundwater recharge, either intentional or unintentional, and through effluent discharge to surface supplies used by other communities, is an accepted practice and will continue until direct reuse itself becomes economical and of accepted safety.

Recreational Use

Recreational water used for body contact sports must be aesthetically acceptable, must not contain substances that are toxic upon ingestion or irritating to the skin, and must be reasonably free from pathogenic organisms. Limits of temperature, color, odor, *p*H, turbidity, and specific ions may be established to define the first two conditions although many state agencies prefer to use qualitative rather than quantitative standards for all but health related conditions. The condition that the water be reasonably free of pathogenic organisms has been subjected to quantitative standards in most states although the standards vary considerably. No definitive relationship between bacterial quality of recreational waters and incidence of related disease has been established. State standards vary from 50–3,000 coliforms/100 ml based either on arithmetical mean, geometrical mean, or the median of monthly samples (*Water Quality Criteria,* McKee, 1963). English studies have concluded that unless the water is so fouled as to be aesthetically revolting, public health standards are reasonably well met. Certainly, these conditions need not be approached in a reservoir made up solely of reclaimed wastewater and the most restrictive state standards of 23 coliforms/100 ml are easily met by the proper combination of the unit processes described in this book.

The suitability of reclaimed water for recreational use has been demonstrated by the Santee County, California, Water District. Treated wastewater is the principal supply for a series of recreational lakes. Secondary effluent from a conventional activated sludge plant passes through an infiltration area to the ponds. Careful monitoring of the bacteriological quality of the infiltration system effluent following chlorination show coliform concentrations of less than 2/100 ml. No virus has been isolated from the water entering the lakes. Based on the high quality of the reclaimed lake waters, the record of community health, and the absence of virus in the lake waters, permission was granted by the State Health Officer to allow establishment of a swimming program during the summer of 1965. The subsequent success of the swimming program demonstrates the ability to treat wastewater to a degree that will meet public health requirements and will be accepted by the public.

In 1969 Indian Creek Reservoir in Alpine County, California, was approved for all water contact sports by all local and state regulatory agencies. This lake is filled with one billion gallons of reclaimed wastewater from the South Tahoe Public Utility District plant at South Lake Tahoe, California. Recreation activities include sailboating, swimming, and trout fishing (see Figure 1–1). The lake has been successfully stocked with rainbow trout. In two summers of sampling, viruses have been isolated on several occasions from the secondary stage of treatment but never from the final chlorinated effluent. Bacteriological tests of water delivered to this reservoir are consistently negative for coliform bacteria.

Figure 1–1 This recreational lake is composed solely of reclaimed water from the South Lake Tahoe, California, advanced wastewater treatment plant. In addition to boating and swimming, a rainbow trout fishery is supported by the lake.

It is difficult to establish separate criteria for noncontact recreation such as boating and aesthetic enjoyment since contact recreation frequently is concurrent with these other activities. Noncontact recreation is adversely affected by problems easily overcome by advanced wastewater treatment; such problems include floating or visible solids, slime growths, algal mats or blooms, discoloration, gas bubbles, turbidity, oil, and foaming. Indeed, it is more likely that a reservoir of renovated wastewater as shown in Figure 1–1 is more in danger of being polluted by boating and shoreline activities than it is in danger of being a source of aesthetic pollution.

Fishing is a recreational activity where water quality is of obvious importance. It is impossible to establish universal quality criteria for fish because the effects of harmful substances vary with species, size, and age of the fish. The effects also vary with the chemical composition of the water supply. Certain salts act synergistically while others act antagonistically. Dissolved oxygen, *p*H, free carbon dioxide, ammonia nitrogen, suspended solids, temperature, and toxic metals are the major parameters to be evaluated for the particular species of fish involved. Warmwater species have been successfully supported in the Santee project reservoirs while rainbow trout have been supported in the South Lake Tahoe effluent reservoir. No health hazards have been related to eating fish from these reservoirs made solely of reclaimed effluent.

Direct water reuse for recreational purposes has been proven practical by the Santee and South Lake Tahoe projects and offers an attractive means of meeting the increased demand for water-based recreation in water-short areas.

UNIT PROCESSES FOR ADVANCED WASTEWATER TREATMENT

A study of the effluent quality produced by conventional secondary treatment processes will quickly reveal that such treatment methods do not remove many pollutants which may create a pollution problem or prevent reuse of the effluent.

Should the presence of materials found in secondary effluent be objectionable due to the desire to reuse the water or the desire to alleviate pollution, the selection from among the appropriate advanced waste treatment unit processes must be made.

Among the many factors to be considered when designing an advanced waste treatment facility are the disposition or use of the final effluent and the related requirements for effluent quality, the nature of the materials to be removed to achieve the required quality, the problems associated with handling of the solids or waste liquids generated in the liquid treatment process, the potential for recovery and reuse of coagulants or other materials used in the treatment processes, the limitations imposed by the sewage collection system and available plant sites, the potential for creating air or land pollution in the process of treating wastewater, and the overall economic feasibility.

The unit processes now being used for advanced waste treatment have generally been used for various industrial purposes and have been adapted to waste treatment plant design as the need for higher effluent quality has developed.

The following sections of this book will explore the design considerations for various unit processes for liquid treatment which have been successfully demonstrated on a plant scale and then discuss the handling of the solids or waste liquids generated by treatment of the liquid phase. This latter point is of major importance since it is obvious that the residues of the waste treatment cannot be discharged into a usable source if a net gain is to be achieved by the advanced waste treatment process. In many instances the disposal of these residues may be the major factor governing the selection of the liquid treatment process.

A later section will briefly discuss other unit processes showing promise but which have yet to be demonstrated beyond the pilot scale.

References

1. Anonymous, "Progress Report of Committee on Quality Tolerances of Water for Industrial Purposes," *Journal New England Water Work Assoc.*, Aug., 1958, p. 1,021.
2. Bouwer, H., "Returning Wastes to the Land, a New Role for Agriculture," *Journal of Soil and Water Conservation*, 23 (Sept.–Oct., 1968).
3. Cillie, G. G., Van Vuuren, L. R. J., Stander, G. J., and Kolbe, F. F,, "The Reclamation of Sewage Effluents for Domestic Use," *Proceedings, Third International Conference on Water Pollution Research*, Munich, Germany. Published by the Water Pollution Control Federation, Washington, D.C., (1966).
4. Connell, C. H., and Forbes, M. C., "Once Used Municipal Water as Industrial Supply, in Retrospect and Prospect," *Water and Sewage Works*, 1964, p. 397.
5. FWQA, *Water Quality Criteria*.
6. Haney, Paul D., "Water Reuse for Public Supply," *Journal American Water Works Assoc.*, 1969, p. 73.

7. Kearns, J. T., "Water Conservation and Its Application to New England," *Journal American Water Works Assoc.* 1966, p. 1379.
8. Merrill, J. C., Jr., and Katko. A., "Reclaimed Waste Water for Santee Recreational Lakes," *Journal Water Pollution Control Federation,* 1966, p. 1310.
9. Metzler, D. F., *et al.*, "Emergency Use of Reclaimed Water for Potable Supply at Chanute, Kansas," *Journal American Water Works Assoc.*, Aug., 1958, p. 1021.
10. Parkhurst, J. D., and Garrison, W. E., "Water Reclamation at Whittier Narrows," *Journal Water Pollution Control Federation,* 1963, p. 1094.
11. Steffen, A. J., "Control of Water Pollution by Waste Utilization: The Role of the WPCF," *Water and Sewage Works,* 1964, p. 384.
12. Stephan, D. G., and Weinberger, L. W., "Water Reuse—Has It Arrived?" *Journal Water Pollution Control Federation,* 1968, p. 529.
13. U.S. Public Health Service, *Draft Policy on Waste Water Reclamation,* unpublished, 1968.
14. *Water Quality Criteria,* edited by J. E. McKee, and J. W. Wolf, Resources Agency of California, State Water Quality Control Board, Pub. No. 3-A (1963).
15. Whetstone, G. A., "Reuse of Effluent in the Future with an Annotated Bibliography," Texas Water Development Board Report 8, Austin, Texas (1965).
16. Wilcox, L. V., "Agricultural Uses of Reclaimed Sewage Effluent," *Sewage Works Journal,* 1948, p. 24.
17. Wolman, A., "Industrial Water Supply From Processed Sewage Treatment Plant Effluent at Baltimore, Md." *Sewage Works Journal,* 1948, p. 15.

2

Chemical Coagulation and Flocculation

GENERAL CONSIDERATIONS

A definition of the difference between coagulation and flocculation is needed because the terms are often used interchangeably. Coagulation involves the reduction of surface charges and the formation of complex hydrous oxides. Coagulation is essentially instantaneous in that the only time required is that necessary for dispersing the chemical coagulants throughout the liquid. Flocculation, which is discussed in a subsequent section, involves the bonding together of the coagulated particles to form settleable or filterable solids by agglomeration. This agglomeration is hastened by stirring the water to increase the collision of coagulated particles. Unlike coagulation, flocculation requires definite time intervals to be accomplished.

The function of chemical coagulation of wastewater may be the removal of suspended solids by destabilization of colloids to increase the settling velocity of settleable material, or removal of soluble inorganic compounds, such as phosphorus, by chemical precipitation or adsorption on chemical floc. The inorganic coagulants commonly used in wastewater coagulation are aluminum salts such as aluminum sulfate (alum), lime, or iron salts such as ferric chloride. Polymeric organic coagulants are also used either as primary coagulants or coagulant aids. Alum and lime coagulation both offer the potential

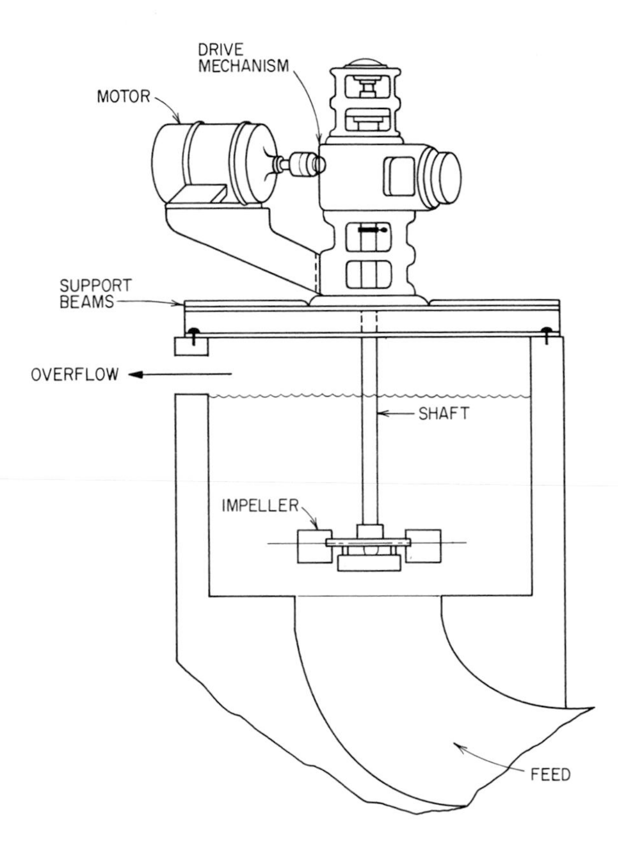

Figure 2–1 Mechanical rapid-mixing device. (*Courtesy Dorr-Oliver*)

of coagulant recovery while no practical means of recovery of iron salts has yet been demonstrated. The sanitary engineering literature contains many discussions of the theory of coagulation, with an excellent summary by O'Melia (1969) available; thus, this section will be limited to practical considerations related to wastewater treatment.

Rapid mixing basins for dispersion of the coagulants are usually equipped with high-speed mixing devices designed to create velocity gradients of 300 fps/ft or more with detention times of 15–60 sec. Power requirements for mechanical mixers are 0.25–1 hp/mgd. A typical rapid mixing mechanism is shown in Figure 2–1.

LIME COAGULATION

Lime reacts with the bicarbonate alkalinity of wastewater to form calcium carbonate and also reacts with orthophosphate to precipitate hydroxyapatite as shown in the following equations:

$$Ca(OH)_2 + Ca(HCO_3)_2 \rightarrow 2\,CaCO_3\downarrow + 2\,H_2O$$
$$5\,Ca + 4\,OH + 3\,HPO_4 \rightarrow Ca_5OH(PO_4)_3\downarrow + 3\,H_2O$$

The apatite precipitate is a crystalline precipitate of variable composition represented by $Ca_5(OH)(PO_4)_3$ in the above equation. Actually, the Ca/P mole ratio may vary from 1.3–2.0. There are several theories proposed in the literature to explain the variable composition of the apatite crystal. It appears likely that the variations result from the substitution of hydrogen ions for calcium ions at the surface and within the crystal (Schmid and McKinney, 1969).

Normally, 70–90 percent of the phosphorus in domestic sewage is in the form of orthophosphate or polyphosphate which may hydrolyze to orthophosphate. Organic-bound phosphorus makes up the rest of the total phosphorus. Although the orthophosphate may be precipitated by calcium ions, Schmid and McKinney (1969) showed that polyphosphate is not readily removed unless orthophosphate is also present, so that the polyphosphate is adsorbed on the floc resulting from the precipitation of the orthophosphate. Phosphorus may also be adsorbed on the surfaces of calcium carbonate particles. The removal of phosphorus upon the addition of lime is achieved by a number of mechanisms operating in parallel with the precipitation of hydroxyapatite shown in the above equation.

Lime is, of course, an alkaline substance which raises the *p*H of the wastewater as it is added. As the *p*H increases above 9.5, precipitation of magnesium hydroxide begins:

$$Mg + Ca(OH)_2 \rightarrow Mg(OH)_2\downarrow + Ca$$

Magnesium precipitation will not be complete until the *p*H reaches 11. Magnesium hydroxide is a gelatinous precipitate which will remove many colloidal solids as it settles. However, its gelatinous properties adversely affect sludge thickening and dewatering.

The solubility of hydroxyapatite decreases rapidly with increasing *p*H, with the result that phosphate removal improves with increasing *p*H. Schmid and McKinney (1969) found that essentially all orthophosphate is converted to an insoluble form at *p*H values above 9.5. They also found that at *p*H values of 9.5 or less phosphorus was adsorbed onto the growing faces of calcium carbonate crystal nuclei, inhibiting growth of the calcium carbonate. As a result, significant quantities of calcium carbonate are not precipitated unless the lime dose is increased to increase the *p*H which will, in turn, decrease the soluble phosphorus concentration while also increasing the driving force of the calcium carbonate precipitation reaction. The practical signifi-

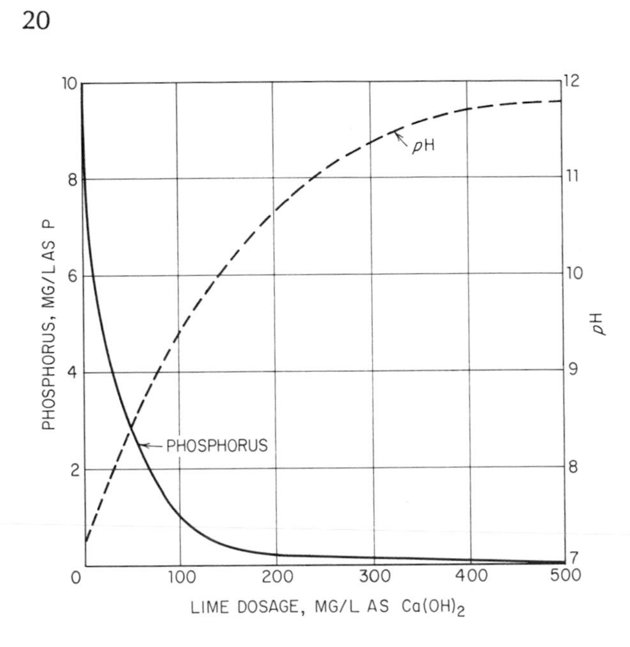

Figure 2–2 Lime coagulation of sewage.

cance of these findings is that lime recalcining may not be practical if the lime dosage does not raise the *p*H to values above 9.5, because calcium carbonate is the only form from which lime can be recovered.

The lime dose required to achieve a given *p*H and turbidity and/or phosphate removal is primarily a function of the wastewater alkalinity and is relatively independent of the influent phosphorus concentration. This is not true where phosphorus removal is obtained with iron or alum salts where an increasing phosphorus concentration requires an increasing amount of iron or aluminum ion. As with any coagulant, laboratory jar tests with a representative sample of the specific wastewater involved enables an accurate evaluation of the coagulation requirements. Figure 2–2 presents results which are representative of the trends obtained with lime coagulation of sewage.

The equipment for storing and feeding of lime and for rapid mixing is essentially the same for wastewater as that which has been used for years in the water treatment field. Lime coagulation can be achieved in a rapid mixing basin with a detention time of 30–60 sec. To maximize phosphorus removal, a fine pinpoint floc should be maintained in the coagulation step. The large surface area of the fine floc will afford more opportunities for phosphorus adsorption on the floc. For this reason, any polymer applied as a settling aid should be added downstream of the rapid mixing and flocculation basins. Experience at the South Lake Tahoe plant (see Figure 13–13) has shown that the additional flocculation provided in a short length of clarifier influent piping from the flocculator to the clarifier is sufficient to produce a large, rapidly settling floc with injection of 0.1–0.5 mg/l polymer, Calgon ST–270, into the flocculator effluent stream. Moving the point of polymer addition from the rapid mixing basin to the flocculator effluent decreased the clarifier effluent phosphorus concentration from 2.5 to 2.0 mg/l at South Lake Tahoe.

Lime is also a term applied to a variety of forms of chemicals which are highly alkaline in character and contain principally calcium and oxygen, but also may contain magnesium. Table 2–1 summarizes information on the various forms of lime. Quicklime, also referred to as burned lime or unslaked lime, is almost entirely calcium oxide produced by calcining limestone at high temperature. High calcium quicklime contains more than 88 percent calcium oxide and less than 5 percent magnesium oxide. Hydrated lime, also known as slaked lime, is calcium hydroxide and is a dry powder obtained by treating quicklime with enough water to satisfy its chemical affinity for water. A high calcium quicklime will produce a hydrated lime with a calcium oxide content of 72–74 percent and 23–24 percent water of hydration. Quicklime and hydrated lime are the forms most often used for wastewater coagulation. Dolomitic lime is obtained by burning dolomitic limestone which is a double carbonate mineral of calcium and magnesium. Dolomitic lime contains from 35–40 percent magnesium oxide. If magnesium content exceeds calcium, the stone is termed dolomitic magnesite. Carbide lime is a waste product from the manufacture of acetylene from calcium carbide and may be competitive with other limes if the point of use is near an acetylene plant. The American Water Works Association has prepared standard specifications for use in water works (AWWA Standard B202-54) which will serve as a useful guide in specifying lime for wastewater coagulation.

As indicated in Table 2–1, lime may be shipped in a variety of containers. In most large plants, bulk shipments of quicklime are preferable. If bag shipment is used, the bags should be stored in a dry room on pallets rather than on concrete floors because moisture can be absorbed through the latter type of floor. Sixty to ninety days is the usual limit for storage in bags because of the gradual adsorption of atmospheric moisture. In small plants where lime consumption is less than a carload a month, hydrated lime is preferred due to ease of handling. In dry storage, deterioration is not serious for periods up to one year.

Lime may be conveyed either mechanically by screw conveyors or bucket conveyors, or pneumatically. Where humidity is high, air slaking of quicklime will take place in pneumatic conveyors due to the absorption of moisture from the atmosphere. Chutes and hopper sides for handling and storing bulk quicklime should have a minimum slope of 55 deg to assure flow since the angle of repose for quicklime is 30–40 deg. Equipment for the mechanical handling of quicklime is generally constructed of mild steel while pneumatic handling equipment is constructed of rubber and steel pipe. Steel bins and standard steel mechanical or air conveying equipment require no precautions against corrosion. Storage facilities for bulk quicklime should be airtight to prevent air slaking. The following table published by the National Lime Association is useful for sizing mechanical screw conveyors:

Table 2-2 Sizing of Screw Conveyors for Lime Transport.

Screw Size (in.)	*Speed (rpm)*	*Capacity (tons/hr)*
6	50	2–2.5
9	50	7–8
12	50	15–20
18	50	45–50

Lime feeders are either volumetric or gravimetric in their means of measuring the quantity of lime added. Gravimetric feeders are more accurate and are generally used in larger plants. In small plants volumetric feeders may be more economical. Both types of feeders can be arranged to feed in proportion to flow of wastewater. Lime is never fed with solution feeders due to its low solubility in water. Lime is not dissolved in the solution pot or dissolver of a dry feeder but rather is merely placed in a slurry form. Quicklime is never applied dry and hydrated lime is rarely applied dry directly to the wastewater because it is more easily transported to the point of application as a slurry, better dispersion of the lime in the wastewater is accomplished, and pre-wetting of the lime in the feeder with violent agitation insures that all particles are wet and that none settle out in the treatment basin. Dissolvers for dry lime feeders are usually designed to provide 3–5 min detention when forming a 6 percent solution (0.5 lb/gal of water). The lime slurry is pumped with open impeller centrifugal pumps usually equipped with iron body and impeller and bronze trim.

LIME SLAKING

The term "slaking" is applied to the addition of water to quicklime or recalcined lime to produce a lime slurry of calcium hydroxide. A lime slaker of either the detention type or pug mill type is used to accomplish slaking in a plant operation. Heat is given off during slaking and the rise in temperature hastens the reaction. Each pound of calcium oxide gives off approximately 490 Btu during slaking. Thermostatic control of either type of slaker is commonly used. The detention slaker produces a creamy slurry of about 10 percent hydrated lime while the pug mill type uses less water for slaking and produces a calcium hydroxide paste containing about 36 percent calcium hydroxide. This paste must be diluted as it leaves the slaking compartment to permit it to flow through pipes to the point of application. When a ratio of 3.5 lb of water to 1 lb of lime is used in detention slakers and the lime contains 90 percent CaO, the temperature of the slaking mixture will rise about 115°F. Pug mill slakers generally use a water-to-lime ratio of 2.5–1. The required slaking time varies with the source of lime. Fast slaking limes will complete the reaction in 3–5 min while poor quality limes may require up to 60 min and an external source of heat. Some investigators have reported that slaking characteristics of lime recovered and recalcined after being used for wastewater coagulation tend to deteriorate. However, this has not been observed in two years of plant operation at South Tahoe. Before selecting a slaker, it is advisable to determine the slaking time, best initial water temperature, and best weight ratio of water-to-lime for the lime to be used, according to the procedures for slaking tests presented by the American Water Works Association. Detention slakers offer some advantages especially when slaking slower slaking limes. Detention-type slakers will slake most limes if proper attention is given to the water-to-lime ratio and temperature, and whether there is hot water or steam available if needed to raise the temperature of a poor-slaking mixture. A vapor removal device is essential to prevent vapor rising into the feeder mechanism or into the slaking room. These devices usually consist of water jets which condense the vapors and wash the dust out of the air before discharge.

An illustrative lime feed system is shown in Figure 2–3. The National Lime Association, Washington, D.C. 20005, can provide information on sources of lime for any specific location as well as more detailed information on handling and feeding of lime.

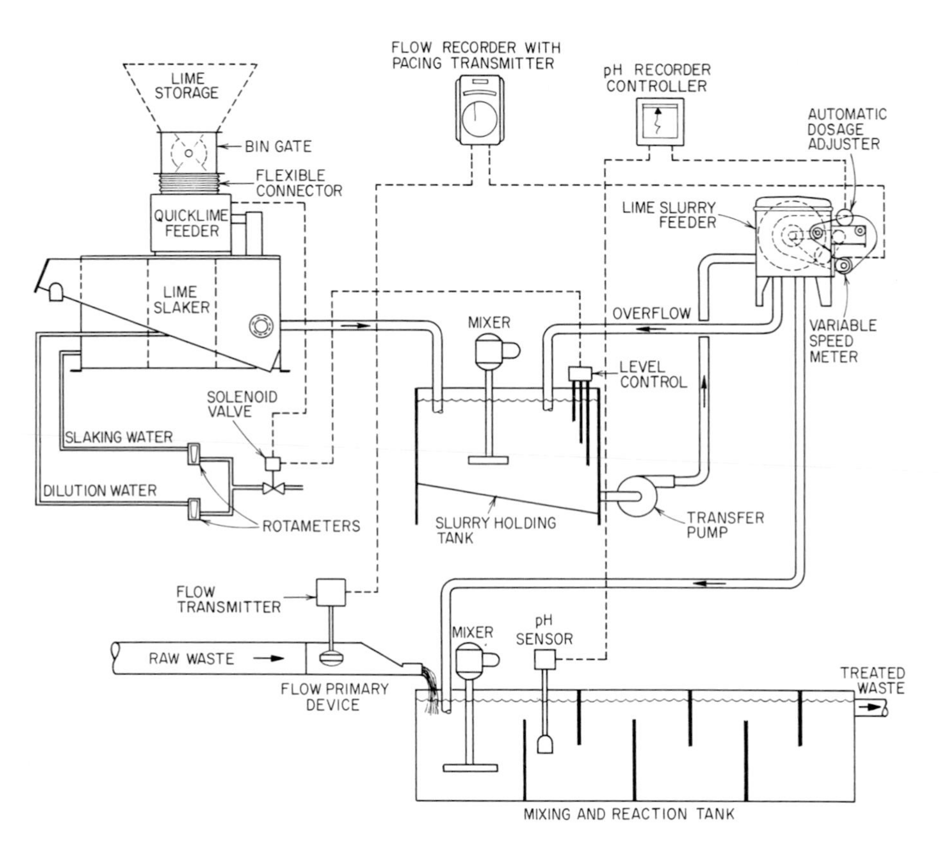

Figure 2–3 Illustrative lime feed system for wastewater coagulation. (*Courtesy BIF Co.*)

ALUMINUM COAGULATION

When alum is added to wastewater in the presence of alkalinity, the following hydrolyzing reaction occurs:

$$Al_2(SO_4)_3 + 6\ HCO_3 \rightarrow 2\ Al(OH)_3\downarrow + 3\ SO_4 + 6\ CO_2$$

This is a classical reaction used in explaining the coagulation process in water treatment plants. The aluminum hydroxide floc is a voluminous, gelatinous floc which enmeshes and adsorbs colloidal particles on the growing floc providing clarification. In the presence of phosphates, the following reaction also occurs:

$$Al_2(SO_4)_3 + 2\ PO_4 \rightarrow 2\ AlPO_4 + 3\ SO_4$$

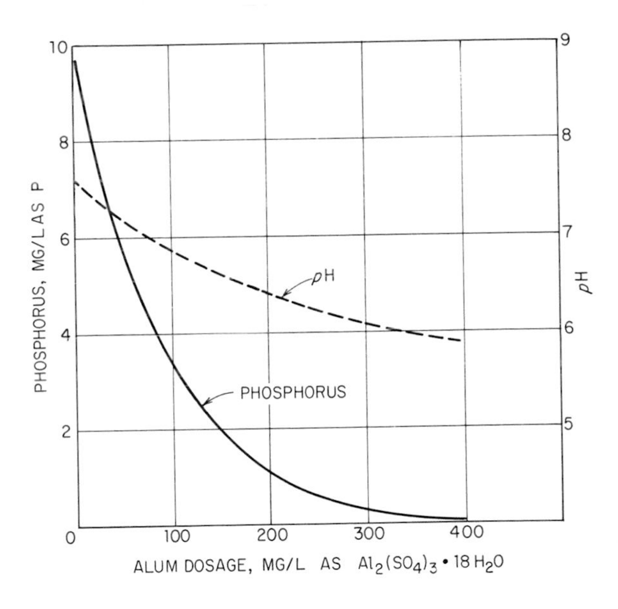

Figure 2–4 Alum coagulation of sewage.

The above two reactions compete for the aluminum ions. At *p*H values above 6.3, the phosphate removal mechanism is either by incorporation in a complex with aluminum or by adsorption on aluminum hydroxide floc. If the hydrolyzing reaction did not compete with the phosphate precipitation reaction, the removal of phosphorus would be stoichiometric, with 0.87 lb of aluminum required for each pound of phosphorus. In practice, the aluminum to phosphorus ratio is on the order 2–3, depending on the final phosphorus concentration desired and the chemical characteristics of the particular wastewater involved. Again, jar tests provide a good indication of chemical requirements. Figure 2–4 illustrates results of alum coagulation.

Sodium aluminate ($NaAlO_2$) is an alternate source of aluminum ion which is usually a poor coagulant in soft wastewaters and a fair-to-good coagulant in hard waters. Parallel jar tests of sodium aluminate and alum will quickly determine if sodium aluminate is an attractive alternate.

Alum is available in either dry or liquid form. It is available in dry form as either ground, powder, or lump form in bags, barrels, or carloads. The ground form is used for dry feeding. It weighs 60–75 lb/ft^3 and is only slightly hygroscopic. The liquid form is a 50 percent alum solution and is usually delivered in minimum loads of 4,000 gal. The liquid form may be delivered by rail or truck. It must be stored and conveyed in corrosion resistant materials such as rubber-lined steel, fiberglass, or stainless steel. Sodium aluminate is also available in either dry or liquid form. The liquid form is normally 43 percent sodium aluminate. Economic merits of dry versus liquid alum will depend primarily on local freight costs. The advantages of the liquid form are the easier handling and feeding.

At South Tahoe, a small dose of alum, in the range of 1–20 mg/l, is always added as a filter aid to the influent of the mixed-media filters (see Figure 13–1). In addition to this one regular point of application, alum may be added if the need arises at several other points in the process, for example in the primary influent, in the rapid mix basin, or in the first-stage recarbonation basin. Alum may be used as the primary coagulant in place of lime or in combination with lime, if desired. For example, if the recarbonation equipment is not operating for *p*H reduction following high lime treatment, then it is possible to feed about 200 mg/l of lime and 100 mg/l of alum and obtain the same treatment of the wastewater without the need for *p*H adjustment by recarbonation. This is done only for short periods of time, however, because the resulting chemical sludge is very difficult to thicken and dewater. In the South Tahoe plant there has been no need to add alum as a coagulant aid after first-stage recarbonation for the calcium carbonate floc settles quite rapidly without the use of alum or other aids. Addition of alum to the primary influent has not proved to be advantageous to date. The alum feed system is depicted in Figure 2–5.

Liquid alum is stored in either or both of two 5,000-gal-capacity tanks. One tank is rubber-lined steel, and the other is made from fiberglass-reinforced polyester resin. There are two alum feed pumps. They are positive displacement metering pumps, each having a capacity of 50.6 gal/hr (25.3 gph for each of two feeder heads). Liquid alum contains about 5.4 lb of dry alum per gallon. Feed can be set over a 10–1 range by adjustment of the length of pump stroke. Dilution water is added to the alum feed pump discharge line to minimize line plugging problems, to reduce delivery time to the point of application, and to assist mixing with the water being treated.

For automatic alum feed control, two flow signals are provided, one of which measures the main plant influent flow, the other measuring the filter effluent flow, which is always substantially greater than plant inflow because of the recycling of various process, cooling, and scrubbing streams within the plant. Either flow signal can be selected for controlling the alum pumps. Ordinarily the dosage is set by the pump stroke adjustment, and the feed is paced to one or the

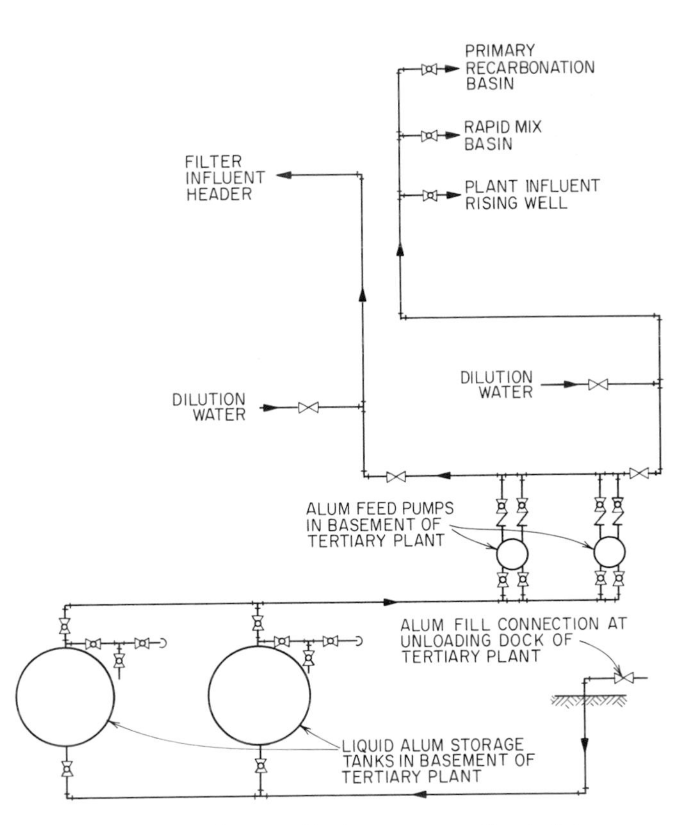

Figure 2–5 Alum application system at South Tahoe.

other of the flow signals depending upon the point of alum application being served.

IRON COAGULATION

The dominant reaction product between the phosphate ion and the ferric ion at *p*H values above 7 is believed to be $FePO_4$ with a solubility product of about 10^{-23} at 25°C (Wuhrmann, 1968). The colloidal particle size of the $FePO_4$ requires a sufficient excess of ferric ion for the formation of a well-flocculating hydroxide precipitate which includes the $FePO_4$ particles and acts as an efficient adsorbent for other phosphorus compounds. Experience has shown that efficient phosphorus removal requires the stoichiometric amount of iron (1.8 mg/l Fe per mg/l of P) to be supplemented by at least 10 mg/l of iron for hydroxide formation. Typically, the iron requirements for municipal wastewaters are 15–30 mg/l as Fe to provide phosphorus reductions of 85–90 percent. The use of 0.3–0.5 mg/l of anionic polymer is required to produce a clear supernatant (Wukasch, 1968).

Ferric chloride is available in dry or liquid form. A crystalline form ($FeCl_3 \cdot 6\ H_2O$) is also available. The crystalline form weighs 60–64 lb/ft^3. The anhydrous form weighs 85–90 lb/ft^3. The typical liquid form contains 35–45 percent ferric chloride and weighs 11.2–12.4 lb/gal. The liquid is handled in rubber, glass, ceramics, and plastic. Ferric sulfate is available in granular form in bags, drums, and bulk, as $Fe_2(SO_4)_3 \cdot 3\ H_2O$ or $Fe_2(SO_4)_3 \cdot 2\ H_2O$. It is normally fed in dry form. The material is hygroscopic and must be stored in tight containers. It weighs 70–72 lb/ft^3. Ferrous sulfate ($FeSO_4 \cdot 7\ H_2O$), also referred to as copperas, is available in granular form in bags, barrels, and bulk. It weighs 63–66 lb/ft^3 and is normally fed dry.

If iron salts are used for wastewater coagulation, a small amount of base, usually sodium hydroxide or lime, is required to neutralize the acidity of these strong acid salts in soft waters. In locales near steel mills, waste pickle liquors which contain large quantities of iron (often about 10 percent iron solutions) may provide an economical source of iron coagulant. The pickle liquors are strongly acid and will normally require that a base be fed to compensate for this acidity.

POLYELECTROLYTES

Polyelectrolytes, or polymers, may be used in advanced wastewater plants as primary coagulants, as flocculation aids, as filter aids, or for sludge conditioning. The discussion at this point will be limited to the use of polymers as coagulants or flocculation aids. Their use for other purposes will be discussed later in connection with the appropriate subject.

In treatment of certain wastewaters, lime or alum alone or in combination may produce a fine or light floc which settles very slowly. If so, there probably is a polymer which, when used as an aid to lime or alum flocculation, will produce a rapidly settling floc. There are certain general guidelines to follow in choosing polymers which may be useful, but the final selection is a matter of trial-and-error. At Tahoe, Purifloc N–11, Calgon ST–270, and others, have been useful as flocculation aids in lime treatment at *p*H 11.0. The usual dose for this purpose is about 0.10–0.25 mg/l of polymer. It is possible to overdose, and this may occur in the rage of 1.0–2.0 mg/l. Addition of the proper amount of polymer at the right point in treatment can greatly improve removals of both turbidity and phosphorus. Jar tests are of some value for preliminary screening, but plant scale tests must be employed for final selection and for determining optimum dosage rates.

Some polymers sold as a dry powder require special procedures for preparation of water solutions. Specific instructions may be obtained from the supplier, but in general the following steps are necessary: thoroughly wet the polymer powder by means of a funnel-type aspirator, add warm water, and mix, usually for about an hour, with gentle, slow stirring until all of the polymer is in solution. Polymer feed solution strengths are usually in the range of 0.2–2.0 percent. Stronger solutions are often too viscous to feed. The solution can be fed by means of positive displacement metering pumps. Typical equipment layouts for polymer mixing and feeding are pictured in the chapter on in-depth filtration.

FLOCCULATION

The purpose of flocculation is to increase the collisions of coagulated solids so that they agglomerate to form settleable or filterable solids. This is accomplished by inducing velocity gradients in the coagulated liquid. The basic equipment is essentially the same as that used for flocculation in water treatment plants for many years. A major difference is that flocculation detention periods may be reduced substantially in wastewater applications in which lime is used as the coagulant. Another major difference is that the goal of chemical addition in wastewater treatment is frequently the removal of soluble inorganics (phosphorus) as well as the removal of collodial solids. As noted in the preceding section, it is actually detrimental to form large floc particles immediately in the flocculation step because it reduces the available floc surface area for adsorption of phosphorus. This is not the case if the goal is strictly one of colloidal solids removal, as it often is in water treatment.

Flocculation may be carried out in a separate basin or in an integral part of the clarifier structure. When carried out in a separate basin, the velocity in the conduits conveying the floc to the clarifier should be 0.5–1.0 fps to prevent disruption of floc. The velocity gradients necessary for flocculation may be induced by mechanical means, such as revolving paddles, or by air diffusion. Camp (1955) and Hud-

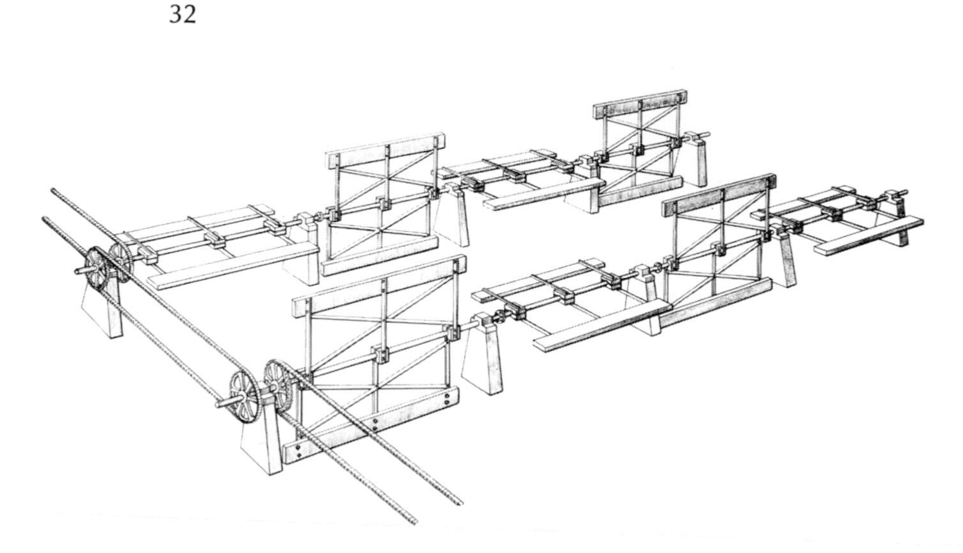

Figure 2–6 Typical paddle-type flocculator. (*Courtesy Dorr-Oliver*)

son (1965) have presented discussions of flocculator design in some detail. In addition, several equipment manufacturers offer design information for flocculator sizing.

Velocities commonly used for flocculator agitators are 0.6 fpm or more with velocity gradients of 30–100 fps/ft. Excessively high velocity gradients will shear floc particles, preventing formation of settleable floc. The use of the highest velocity gradient compatible with the strength of the floc involved will help to produce compact floc particles. Flocculator detention times of 15–60 min have been commonly used in water treatment plant design.

Rose (1968) reports that the sludge blanket flocculation used in many water clarifiers is of questionable use in sewage applications. In this type of clarifier, the solids contact provided as the incoming solids flow through a blanket of solids maintained in the clarifier provides flocculation. However, Rose found that in wastewater treatment it is undesirable to build up a blanket because the sludge contains organics and may turn anaerobic rapidly, impeding phosphorus removal and clarification. Similar difficulties with solids contact units have been reported at the Washington, D.C., pilot plant.

Experience at the South Lake Tahoe plant has shown that flocculation times as low as five minutes are adequate with lime coagulation.

Lime Coagulation and Flocculation System at South Tahoe

Facilities for adding up to 400 mg/l of CaO to a maximum flow of 7.5 mgd are installed at Tahoe. Duplicate equipment is provided for storing, feeding, and slaking commercial quicklime and in-plant recalcined quicklime.

Fresh, or makeup, lime is unloaded in the bulk from delivery trucks using a pneumatic conveying system which has a rated capacity of 10 tons/hr and is discharged into a 35-ton-capacity overhead steel storage bin.

Recalcined lime is conveyed from the lime furnace to a second 35-ton bin by a separate pneumatic conveying system having a capacity of 0.75 ton/hr.

Mechanical devices such as bucket elevators and screw conveyors could be used rather than the pneumatic systems, but they create severe dust problems not associated with the pneumatic conveyors. However, mechanical conveyors are easier to operate and maintain.

The piping to the lime bins is arranged so that either bin may be used to receive fresh or recalcined lime, if the need arises. Each overhead lime storage bin is located directly above a gravimetric belt-type lime feeder, which discharges to a paste-type lime slaker. Each unit has a capacity of 1,500 lb of CaO per hour. Ordinarily, about 25 percent fresh makeup lime and 75 percent recalcined lime is fed so as to maintain a *p*H = 11.0 in the rapid mix basin. For the Tahoe wastewater, this requires a total of about 400 mg/l of lime. In other wastewaters, as little as 100 mg/l of lime may provide the desired phosphorus removal. The only way that the required lime dosage can be determined exactly for a particular situation is by trial and error, either in jar tests or in plant operation, using the actual wastewater to be treated.

At Tahoe, separate basins are provided for rapid mixing, flocculation, and settling. For municipal wastewater, this is usually the best arrangement, rather than to attempt to combine all three functions into a single basin, because it provides complete control over each process individually. The solids contact provided by exterior recirculation of settled lime floc, is often desirable. For several reasons, this is much more practical than internal retention of these solids in a solids contact unit. In the event of heavy carry-over of organic solids from the pretreatment process, the return of settled lime solids can be immediately discontinued, thus avoiding a serious upset of the chemical clarification process which would unavoidably occur in a solids contact unit. It is often impossible to maintain a sludge blanket in a solids contact unit in wastewater treatment because of the rapid changes in influent water composition which are characteristic of sewage. Also, the organic content of the lime sludge often precludes holding a blanket at the proper level in the basin, as discussed in Chapter 3. Separate mixing, flocculation, and settling basins also provide more flexibility in selecting points of application for different chemicals, especially polymers. Plant operating experience has shown conclusively that it is highly beneficial to withdraw the settled lime floc from the clarifier as rapidly as possible rather than to hold it in the basin.

Ideally, the lime slurry from the slakers would be discharged directly into a rapid mix basin located directly beneath the slakers since lime slurry plugs pipelines rather rapidly. Unfortunately, the rapid mix basin frequently must be located 100–300 ft away from the slakers. At Tahoe, a 6 in. asbestos-cement pipeline is used to convey the lime slurry by gravity flow for a distance of 300 ft to the rapid mix basin.

This line is cleaned once each week by use of a high-pressure hydraulic sewer cleaner. A series of manholes on the lime feed line located about 50 ft apart provide easy access for this cleaning operation. The original feed line consisted of a 3-in. diameter rubber hose placed within the 6-in. asbestos-cement carrier line. Two complete sets of hose with quick-couplings located at each manhole were provided, which allowed each length of hose to be removed for cleaning as it was replaced with the clean duplicate length. This was a satisfactory arrangement, but less cleaning time is required using the hydro-cleaning of the larger line.

The rapid mix basin provides about one minute of mixing at design flow. A vertical shaft mechanical mixer is used. Lime deposits build up on the metal mixer paddles, which require frequent cleaning to avoid shaft vibration. In the rapid mix basin, not only lime, but also alum, polymer, or recycled settled lime solids may be added. Ordinarily no polymer is added in the rapid mix, but rather it is added between the flocculation basin and the chemical clarifier basin for better phosporus adsorption on the increased surface area provided by the finer floc particles resulting from delayed polymer addition.

Following rapid mixing, five minutes of flocculation is provided by air agitation, if desired. However, experience in plant operation indicates that better treatment is obtained without this additional mixing.

Lime feed rates are usually controlled by manual setting of the gravimetric feeders. This manual control has proven quite satisfactory. Automatic control is also available, but is seldom used. The automatic control system consists of a *p*H electrode in the rapid mix basin with a signal transmitted to an indicating-recording-control unit located near the lime feeders. This *p*H signal can be used to control the lime feed rate to maintain any desired *p*H in the rapid mix basin by making the appropriate setting on the controller. This control works satisfactorily except that the *p*H in the rapid mix basin is always just slightly above or below the set point and never exactly on it except for a very short time, due to time lag in the control and feed system. The operators prefer to use the manual control.

References

1. Camp, T. R., Flocculation and Flocculation Basins," *Transactions American Society of Civil Engineers,* 120:1 (1955).
2. ———, "Floc Volume Concentration," *Journal American Water Works Assoc.* 1968, p. 656.
3. *Chemical Feeder Guide,* BIF Co., Providence, Rhode Island, 1969.
4. Hudson, H. E., "Physical Aspects of Flocculation," *Journal American Water Works Association,* 1965, p. 855.
5. Morgan, J. J., and Englebrecht, R. S., "Effects of Phosphates on Coagulation and Sedimentation of Turbid Waters," *Journal American Water Works Assoc.,* 1960, p. 303.
6. Mulbarger, M. C., *et al.,* "Lime Clarification, Recovery, Reuse, and Sludge De-Watering Characteristics," *Journal Water Pollution Control Federation,* 1969, p. 2070.
7. O'Melia, C. R., "A Review of the Coagulation Process," *Public Works,* May, 1969, p. 87.
8. Rose, J. L., "Removal of Phosphorus By Alum." Presented at the FWPCA Seminar on Phosphate Removal, Chicago, Illinois (June 26, 1968).
9. Schmid, L. A., and McKinney, R. E., "Phosphate Removal by a Lime-Biological Treatment Scheme," *Journal Water Pollution Control Federation,* 1969, p. 1259.
10. Stumm, W., and Morgan, J. J., "Chemical Aspects of Coagulation," *Journal American Water Works Assoc.,* 1962, p. 971.
11. Walker, J. D., "High Energy Flocculation," *Journal American Water Works Assoc.,* 1968, p. 1271.
12. Wuhrmann, K., "Objectives, Technology, and Results of Nitrogen and Phosphorus Removal," *Advances in Water Quality Improvement, I,* University of Texas Press, Austin, Texas, 1968.
13. Wukasch, R. F., "The Dow Process for Phosphorus Removal." Presented at FWPCA Seminar on Phosphate Removal, Chicago, Illinois (June 26, 1968).

3

Sedimentation

IMPORTANCE OF SEDIMENTATION BASIN DESIGN

The importance of adequate sedimentation basin design is a point not to be taken lightly in the design of an advanced waste treatment plant. A "cookbook" approach using overflow rates accepted as standard criteria for other applications for a particular type of clarifier equipment may lead to poor or even totally inadequate plant performance. In many cases, the apparently simple process of gravity separation of settleable solids may be the weak link in the chain of treatment processes on which relatively complex downstream processes are dependent. The analyses of the capital and operating costs of advanced waste treatment plants presented elsewhere in this book show that the capital cost of the clarifier equipment is indeed a very small part of the overall treatment costs. Any attempt to economize in clarifier design offers little potential for significant overall dollar savings but much potential for destroying plant reliability. If an immediately downstream filter or adsorption process cannot tolerate high solids loadings, gross carry-over of solids from the clarifier may cause the entire plant to be removed from service. Situations which require advanced waste treatment are rarely suited for periods of plants bypass. Although the plant designer may find selection of clarifier design criteria to appear to be a relatively mundane task compared to delving into some unit processes new to him as a waste treatment plant designer, he will be wise to devote a significant effort to determination of the criteria applicable to his particular problem.

CONVENTIONAL CLARIFIERS

Among the general types of clarifiers available for gravity sedimentation of settleable solids are the following:

1. Horizontal flow basins with external flocculation.
2. Horizontal flow basins with internal flocculation (usually circular).
3. Upflow basins with internal rapid mixing and flocculation (usually circular.)
4. Solids contact units with recirculation of settled solids.

Several manufacturers offer equipment in the above categories with several variations on basin inlet and outlet systems, recirculation of the sludge blanket, and control of the sludge blanket level available. Detailed literature describing each type of equipment is readily available from manufacturers and no attempt will be made here to review all available equipment. Rather, the experiences gained in field application of the general clarifier types available for advanced waste treatment processes will be examined. Only applications will be included which involve separation of floc resulting from chemical coagulation of sewage, since performance of conventional primary and secondary sewage clarifiers has been reviewed in detail in other publications (ASCE Manual No. 36).

Most of the field experience indicates that surface overflow rates recommended for application of specific equipment to removal of chemical floc in water treatment applications must be reduced for removal of floc from chemically coagulated sewage. Also, maintenance of sludge blankets in upflow basins has proven difficult in these applications.

Rose (1968) reports that an overflow rate of 720 gpd/ft^2 should not be exceeded for the removal of alum floc from coagulated secondary effluent. Rose evaluated a circular clarifier with internal flocculation equipment normally used as a sludge blanket unit of the solids contact type in water treatment applications at overflow rates in excess of 1400 gpd/ft^2. However, Rose found that halving of the rate was required and that maintenance of a sludge blanket was not feasible. He reports the alum floc in sewage to be quite light and fragile and that polymers may be required as settling aids for operation even at 720 gpd/ft^2 to prevent massive overflow of floc. Rose emphasizes the importance of removing sludge at a rate adequate to prevent septicity of the sludge from occurring. He found that a sludge removal rate of 3–5 percent of the throughput was required to prevent septicity.

O'Farrell, *et al.* (1969), also report occasional difficulty with a solids contact type of clarifier applied to lime-coagulated secondary effluent. They report the slurry pool in the clarifier was often unstable and expanded into a sludge blanket which would periodically overflow the clarifier. Septic solids occasionally forced shutdown of the clarifier. The chief difference between stable and unstable sludge blankets was reported by O'Farrell, *et al.*, to be the ratio of calcium carbonate to noncarbonate solids in sludge. Calcium carbonate concentrations in excess of 75 percent maintained a stable slurry pool. A sludge withdrawal rate of 0.5 percent of the total flow was adequate with a stable slurry, and a sludge concentration of 10–12 percent solids was maintained. With a stable slurry, the clarifier produced effluent turbidities of 3–6.5 Jackson units (JU) and a phosphorus concentration of about 0.5 mg/l as P at an overflow rate of 1,500 gpd/ft^2 with a water temperature of 25°C. A decrease in wastewater temperature to 17°C caused the effluent turbidity to increase to 16 JU and the phosphorus to 1.8 mg/l at the same overflow rate. The quality of the secondary effluent being coagulated and settled also worsened at the lower temperature.

Convery (1968) recommends an overflow rate of 700 gpd/ft^2 for removal of alum floc from coagulated sewage. He reports that basin performance continued to improve as the rate was decreased to 300 gpd/ft^2 but that a rate of 700 gpd/ft^2 represented the best balance between basin performance and economic considerations. Weber, *et al.* (1970) also report successful operations at 700 gpd/ft^2 for separation of lime floc from coagulated raw sewage. Duff, *et al.* (1968) report success in a 13 mgd solids contact plant at an overflow rate of 1,020 gpd/ft^2 using lime as the primary coagulant with some alum being used. Data presented by Kalinske, *et al.* (1968) also confirm that higher overflow rates are permissible when using lime rather than alum. They suggest a design rate (solids contact unit) of 1,800 gpd/ft^2 when using lime and 1,200 gpd/ft^2 when using alum for coagulation of secondary effluents. Substantial polymer doses (0.25 mg/l) were used as settling aids in order to achieve these rates in a 25-gpm pilot unit. A rate of 1,400 gpd/ft^2 was suggested for lime coagulated raw sewage. Plant-scale experiences of others indicate that these overflow rates may be difficult to consistently maintain in a large-scale clarifier.

The cited experiences and other unpublished experiences indicate that maintenance of sludge blankets in clarifiers treating chemically coagulated sewage creates an unstable system which is easily upset and cannot consistently provide the degree of reliability which should be associated with an advanced waste treatment plant. It appears that surface overflow rates of 700–900 gpd/ft^2 should be used in the absence of sludge blankets.

Provision should be made for recirculation of controlled amounts of sludge from the bottom of the clarifier to the rapid mix inlet. The high *p*H of lime-treated water will form deposits of calcium carbonate on structures and in pipelines which it contacts. Lime sludge suction lines should be glass lined to facilitate cleaning. Provisions must also be made for regular cleaning of all other pipelines which carry the high *p*H effluent. Use of the new polyurethane cleaning pigs should be compatible with the layout of the pipelines. Mechanical sludge collection equipment used in lime settling basins should be of the bottom scraper type rather than the vacuum pickup style because of the extremely dense sludge to be handled.

The chemical clarifier at the South Tahoe Water Reclamation Plant is designed for wastewater flows of 7½ mgd. The influent water has been flocculated with about 400 mg/l of lime as CaO to *p*H = 11, and about 0.25 mg/l of a polyelectrolyte (Calgon ST-270 or Dow N-11). The basin is a conventional center inlet circular basin (Dorr-Oliver) 100 ft in diameter by 10-ft side water depth. Basin overflow rate at design flow is 950 gpd/ft^2, and the weir rate is 24,000 gpd/-linear ft. The settled lime sludge withdrawn from the basin contains 0.5–2.0 percent solids. There are two sludge pumps. One is a variable-speed drive centrifugal pump rated at 500 gpm. The other is a positive displacement pump (Moyno) with a capacity of 100 gpm. Both pumps discharge to a lime mud thickener, or, alternately, to the primary clarifier. About 30 gpm of sludge is recycled to the rapid mix basin. The lime mud suction line beneath the concrete bottom of the clarifier is glass-lined pipe. The discharge line is cast-iron pipe which is cleaned about once a year by the use of a polyurethane pig. The clarifier effluent turbidity is in the range of 2–20 JU under a wide variety of hydraulic and solids loadings.

SHALLOW DEPTH SETTLING DEVICES

A considerable amount of interest in applying the long-recognized principles of shallow depth settling devices has developed recently. A paper by Hansen and Culp (1967) presented the first recent dis-

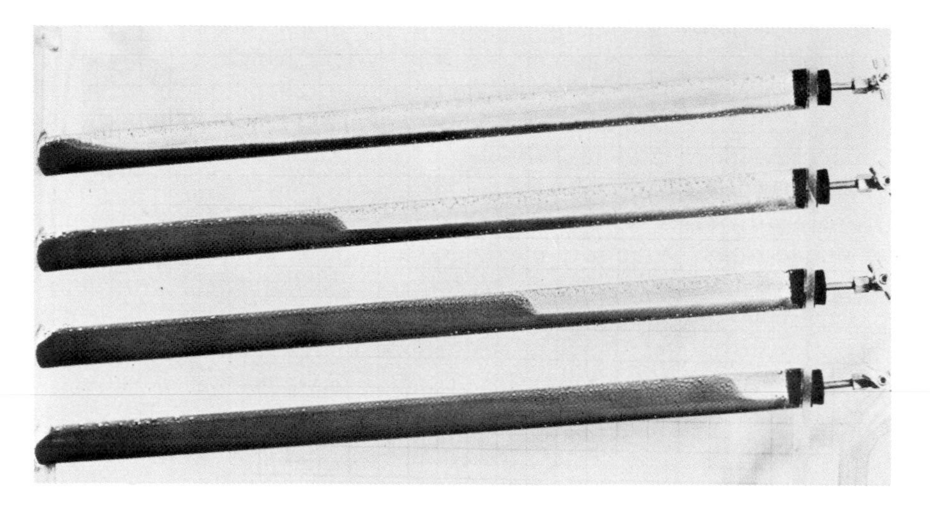

Figure 3–1 Demonstration of essentially horizontal tube settler. The bottom tube is almost full of sludge and ready for draining. The other tubes were placed in service at later intervals to show the progressive filling of tubes with sludge. (*Courtesy Neptune Microfloc, Inc.*)

cussion of attempts to make practical application of the principles first proposed by Hazen (1904). Hansen and Culp (1967) reported that passing the flow longitudinally through small diameter (1–2 in.) horizontal tubes resulted in efficient removal of settleable solids with settling compartment detention times on the order of 5–10 min. The settled solids accumulate on the bottom of the horizontal tubes until the tube is essentially full of sludge. The tubes shown in Figure 3-1 illustrate the progressive filling with sludge. These tubes were 1 in. in diameter and 2 ft long with flow passing from left to right at a rate of 5 gpm/ft^2 of tube cross-sectional area. The tube detention time was 3 min. The accumulated sludge is removed from the horizontal tubes by draining them. In practice, this draining operation is coordinated with that of a filter following the tube settler. Each time the filter backwashes, the tube settler is completely drained. The tubes are inclined upward only slightly in the direction of normal flow (5–7 deg) to promote the draining of sludge during the backwash cycle. The rapidly falling water surface scours the sludge deposits from the tubes and carries them to waste. The water drained from the tubes is replaced with the last portion of the filter backwash water, thus no additional water is lost due to the tube draining procedure and no mechanical sludge removal equipment is required.

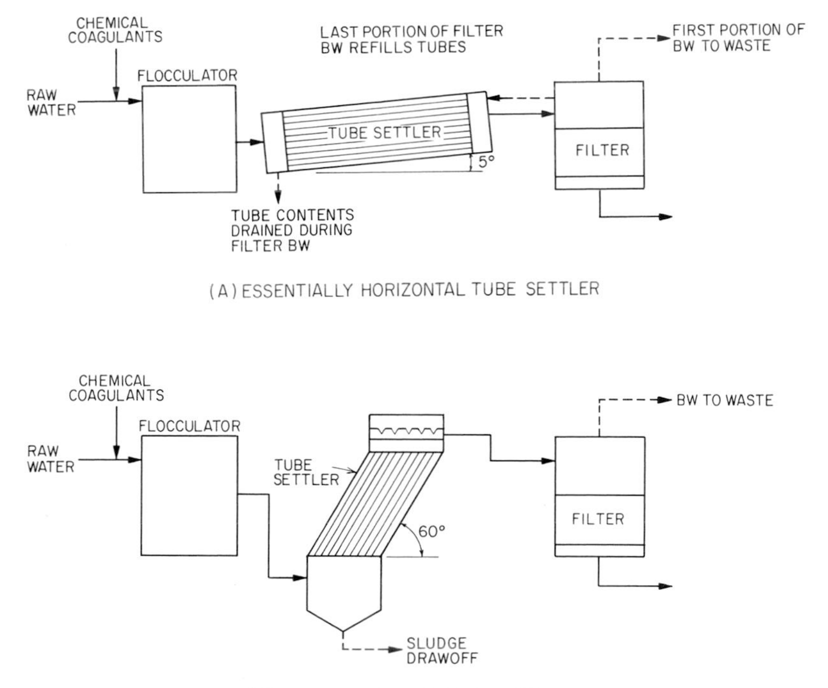

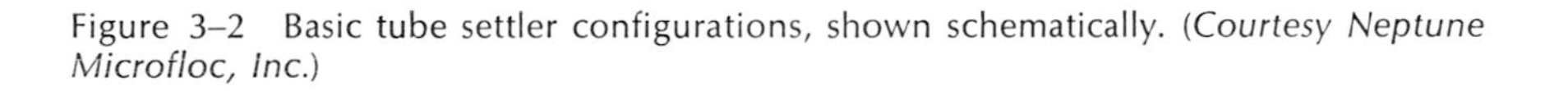
Figure 3–2 Basic tube settler configurations, shown schematically. (*Courtesy Neptune Microfloc, Inc.*)

Later papers by Culp, Hansen, and Richardson (1968) and Hansen, Culp, and Stukenberg (1969) described the development of a "steeply inclined" tube settler. They found that sediment in tubes inclined at angles in excess of 45 deg would not accumulate but would move down the tube to eventually exit the tubes into the plenum below. This, of course, eliminated the need to drain the tube for sludge removal. Figure 3-2 illustrates the two tube settler concepts now being marketed in the United States. The theoretical basis of the tube settling concept has been discussed recently by McMichael (1969) and Yao (1970). There is general agreement that the tube settler concept offers a theoretically sound basis for operating clarifiers at surface loading rates substantially (2–4 times) higher than in deep, conventional basins.

Culp, Hsiung, and Conley (1969) have presented a summary of

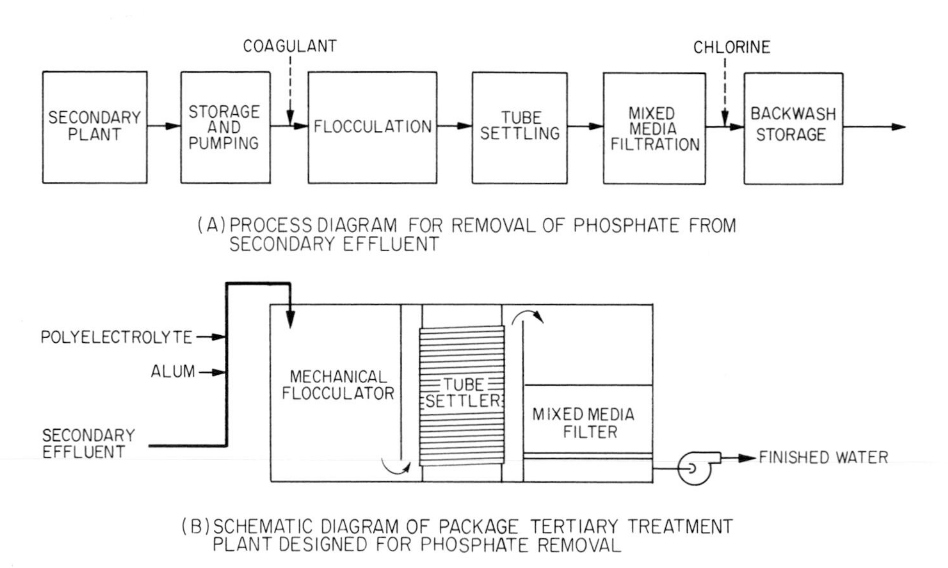

Figure 3–3 Package plant system for phosphate removal using tube settling concept. *(Courtesy Neptune Microfloc, Inc.)*

operating experiences with tube settling. Most of the plant-scale experience reported deals with water treatment applications rather than waste treatment applications. Data were presented on specific application of the horizontal tubes in a system of the type shown in Figure 3-3 for removal of turbidity and phosphate from secondary effluent. Final effluent turbidities of 0.3–0.7 JU and phosphates of less than 0.5 mg/l as PO_4 were reported. However, the physical limitations imposed by the requirement that the tubes be drained for sludge removal limits the practical capacity to plants of 1 mgd or less in capacity. Package steel plants are available for tertiary plants in this capacity range. Their compact size appears to offer some substantial savings worthy of consideration by a design engineer faced with new plant design. The tertiary package plant is generally operated at a constant flow rate. Usually the provision of surge storage to accept peak flow above the average operating rate of the tertiary plant is more economical than sizing the tertiary facility to handle the peak flows. An example of the calculation of surge storage requirement is presented in Table 3-1. The filter backwash wastewaters are returned to the storage pond so that they may be reprocessed in the tertiary plant. The chemical sludge from the tube settlers is generally drained to a sludge storage tank prior to disposal or recovery. A 25,000 gpd package plant of the type shown in Figure 3-3

Table 3-1 Calculation of Surge Storage Requirements. *(Courtesy Neptune Microfloc, Inc.)*

Daily sewage flow = 200,000 gpd

Tertiary plant: flow = 200 gpm (12,000 gph), backwash volume = 6,800 gallons

	POND INFLUENT				ACCUMULATION	TOTAL
Time	*Raw sewage gallons*	*Backwash return gallons*	*Total gallons*	POND EFFLUENT *gallons*	IN POND *gallons*	BUILDUP IN POND *gallons*
6–7 AM	6,000	0	6,000	12,000	0	0
7–8	3,000	0	3,000	12,000	0	0
8–9	8,000	0	8,000	12,000	0	0
9–10	9,000	0	9,000	12,000	0	0
10–11	13,000	6,800	19,800	9,000	10,800	10,800
11–12 Noon	13,000	0	13,000	12,000	1,000	11,800
12–1 PM	13,000	6,800	19,800	9,000	10,800	22,600
1–2	13,000	0	13,000	12,000	1,000	23,600
2–3	9,000	6,800	15,800	9,000	6,800	30,400
3–4	9,000	0	9,000	12,000	−3,000	27,400
4–5	9,000	6,800	15,800	9,000	6,800	34,200
5–6	9,000	0	9,000	12,000	−3,000	31,200
6–7	10,000	0	10,000	12,000	−2,000	29,200
7–8	10,000	0	10,000	12,000	−2,000	27,200
8–9	10,000	0	10,000	12,000	−2,000	25,200
9–10	11,000	0	11,000	12,000	−1,000	24,200
10–11	10,000	0	10,000	12,000	−2,000	22,200
11–12 Midnight	10,000	0	10,000	12,000	−2,000	20,200
12–1 AM	3,000	0	3,000	12,000	−9,000	11,200
1–2	3,000	0	3,000	12,000	−9,000	2,200
2–3	3,000	0	3,000	12,000	−9,000	0
3–4	3,000	0	3,000	12,000	0	0
4–5	5,000	6,800	11,800	9,000	2,800	2,800
5–6	5,000	0	5,000	12,000	−7,000	0

Operating volume needed in holding pond = 34,200 gallons.

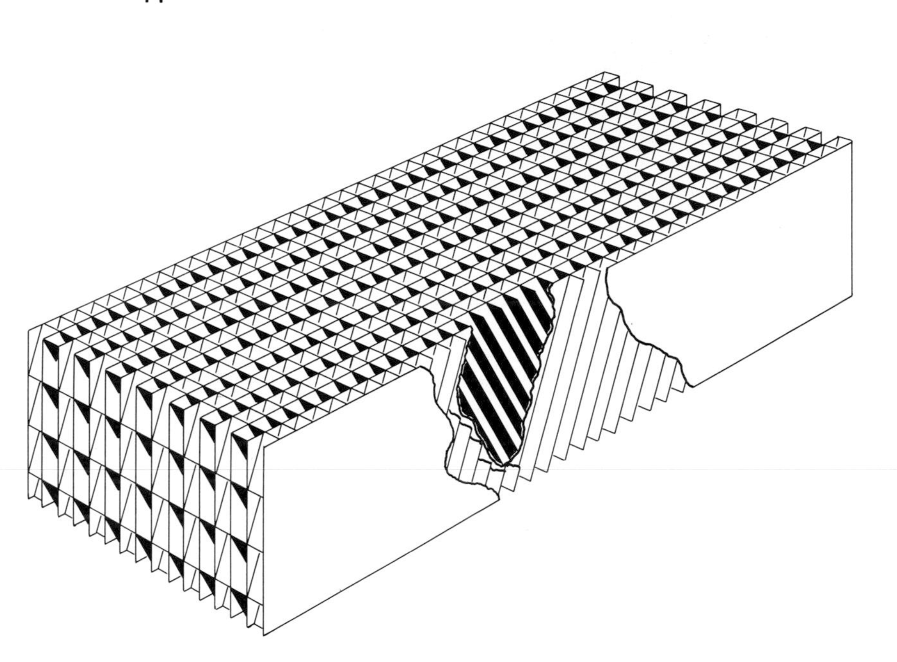

Figure 3–4 Module of steeply inclined tubes. (*Courtesy Neptune Microfloc, Inc.*)

providing flocculation, settling, and filtration has dimensions of only 5 × 5 × 6 ft high.

The greatest potential for application of the shallow depth settling principle lies in the use of the steeply inclined tube settler for the primary, secondary, and tertiary clarification steps. A design (Figure 3-4) has been developed which incorporates a multitude of steeply inclined 2-in. square tubes in an all-plastic tube module. Extruded plastic channels are installed at a 60 deg inclination between thin sheets of plastic, normally PVC. By inclining the tube passageways in alternate directions, to provide rigidity, a rectangular module that can be readily installed in either rectangular or circular basins can be fabricated. Typical module dimensions are 10 ft long × 2 ft 6 in. wide. Thus, the large amount of tube settling area required for plants of large capacity can readily be installed. At the time of this writing, the plant-scale experience in steeply inclined clarification of sewage is limited, which makes presentation of well-defined design criteria impossible. The potential merits are illustrated in a paper by Livingston (1969) which describes a water treatment plant (Buffalo Pond, Saskatchewan) clarifier employing tubes to provide an increase in the clarifier capacity of 3.0 mgd as a conventional basin to 7.5 mgd (about 2 gpm/ft^2 of tube area) when equipped with tubes. The average effluent turbidity of a parallel conventional basin operating at 3 mgd was 0.7 JU while the basin equipped with tubes produced an average effluent turbidity of 0.6 JU. Thus, a substantial increase in capacity was achieved at no sacrifice in effluent quality. However, floc deposition occurred at the top end of the tubes. Culp, *et al.* (1969) reported the floc buildup accumulated as deeply as 10 in. before it was removed by momentarily dropping the water level below the bottom of the tubes to remove the buildup. The frequency of cleaning by this method varied from once in two months to once a week. The plant personnel later installed a series of submerged jets, similar to a fixed surface wash in a filter, just above the upper tube surface. Intermittent, daily operation of the horizontal jets has proved effective in keeping the upper tube surface clean. Similar buildup of floc on the upper module surface has been reported from several other installations. In water treatment plants, the accumulation of inorganic materials has presented no major operating problems as evidenced by the fact that the Buffalo Pond plant discussed above has recently converted all of its remaining clarifiers to the tube concept following a lengthy evaluation of one basin with tubes. However, a buildup of organic materials in a wastewater application could cause degradation of effluent quality should the solids turn septic and float to the surface. A suitable cleaning system would overcome this potential problem. The preliminary results from small-scale wastewater installations of steeply inclined tubes are encouraging. Culp, *et al.* (1969) report efficient activated sludge clarification at rates of 2 gpm/ft^2 with 4-ft-long tubes and efficient primary clarification at 3 gpm/ft^2 with 2-ft-long tubes. Surface buildup of solids were observed in both types of application but had no apparent adverse effect on performance provided adequate cleaning techniques were used. Bologna (1969) also reported the successful application of tubes to activated sludge clarification but felt that the upper loading limit was 1 gpm/ft^2, considerably lower than that reported by the manufacturer. It appears that more plant-scale experience will be required before definitive design criteria can be presented for applying tube settlers to primary and secondary wastewater clarification. The results to date certainly indicate the concept has enough potential that a designer should carefully explore the applicability of the concept in light of experiences subsequent to this writing before designing any new clarifier.

The rather extensive experience with the steeply inclined tubes in treatment of coagulated waters indicates that substantial economies may be realized in using the concept for clarification of coagu-

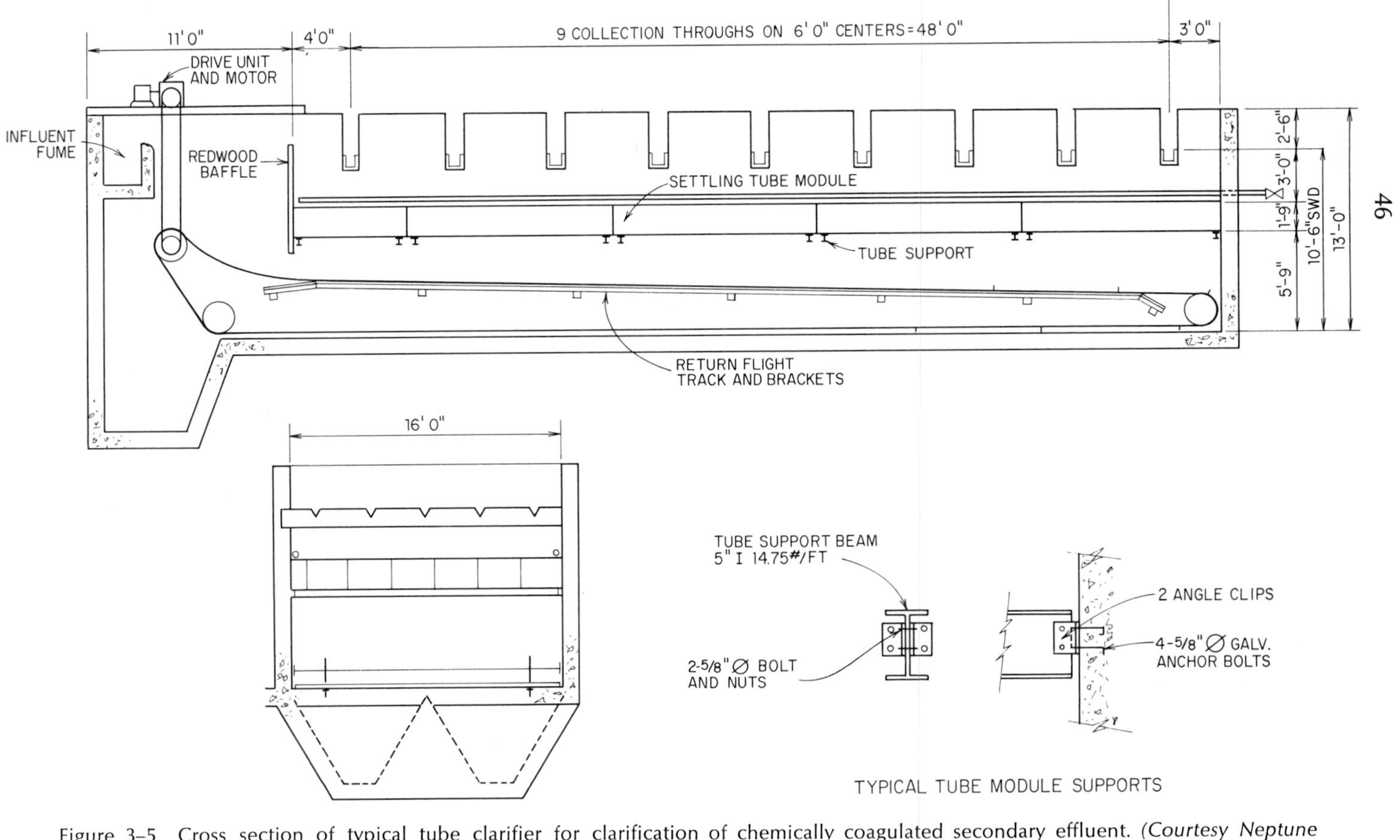

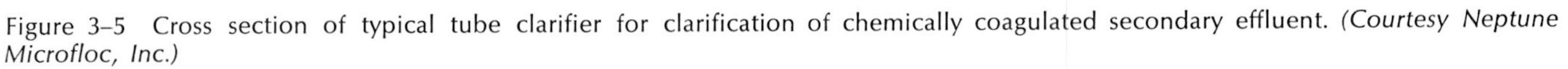
Figure 3–5 Cross section of typical tube clarifier for clarification of chemically coagulated secondary effluent. *(Courtesy Neptune Microfloc, Inc.)*

lated secondary effluents at loading rates of about 2 gpm/ft^2 of tube entrance area. Pilot tests on coagulated secondary effluent indicate that higher rates may be possible. Figure 3-5 presents a typical cross section of a rectangular clarifier design used for removal of chemical floc. The extensive effluent launder system is designed to insure adequate flow distribution through the tubes. With the exception of the added launders and the need for tube supports, the basin design would follow conventional practice. Of course, the sludge removal system would have to be sized to handle the greater quantity of sludge, than would occur in a conventional basin of the same size.

SUPPLEMENTAL SETTLING FOR TERTIARY TREATMENT

Plain settling (settling without chemical coagulation) of a secondary effluent in ponds or basins has been suggested as a method of improving secondary effluent BOD, suspended solids, and coliform concentrations most frequently in conjunction with small package plants. Some states require "polishing" ponds downstream of all package sewage treatment plants to intercept the slug discharge of solids which have been known to occur frequently from these small plants. The difficulty in design of these ponds is the selection of a detention time which will be long enough to remove the settleable material and to balance the sludge accumulation rate with the ability of the pond to stabilize the sludge but short enough to minimize algal growths. The abundance of algal nutrients in secondary effluents coupled with the relatively low turbidity of a good effluent can lead to profuse algal growths which would result in an increase rather than a reduction in organic and suspended solids discharged from the treatment plant. A study by Morris, *et al.* (1963) indicated that detention times of more than 3 days could result in profuse algal growths. The algal growth with a 5-day detention time was found to increase the solids and organic loading on the receiving stream except during periods of very heavy solids losses from the secondary plant. A reduction in the detention time from 5–3 days reduced the algal concentration by a factor of four and appeared to provide adequate sludge stabilization capacity for moderate losses of solids from the secondary plant. The 3-day detention provided an 80 percent reduction in coliform bacteria. A 1-day detention pond was found to provide inadequate sludge storage capacity and became anaerobic following a period of settleable solid loss from the secondary plant. Oakley and Cripps (1969) reported the operation of

Figure 3–6 Tube modules being installed in rectangular clarifier. (*Courtesy Neptune Microfloc, Inc.*)

Figure 3–7 Settling tube module being placed on support beam. (*Courtesy Neptune Microfloc, Inc.*)

the Rye Meads Works (England) effluent lagoons. Three lagoons in series provided a total retention of 17 days of secondary plant effluent. The lagoons reduced the suspended solids of the secondary effluent from a range of 8–53 mg/l to 0.7–9.3 mg/l and the biochemical oxygen demand (BOD) from 3.5–38 mg/1 to 1.4–6.9 mg/l. The potential problems of most concern were those associated with the growth of algae in the lagoons and the possible effects of these if they passed into the river. The difficulty of sludge clearance after a number of years was also listed as a potential problem. However, over the short term, the lagoons were providing a substantial improvement in effluent quality.

A British publication (Anonymous, 1967) concludes that polishing pond detentions of 3–4 days is optimum and that several polishing ponds in series should be used to minimize the loss of floating solids. They conclude that the performance of such ponds is only moderate except in regard to removal of coliform bacteria.

It appears that the application of plain settling in a polishing pond is limited in application to small package sewage plants where an average effluent quality no better than secondary quality is acceptable, where the moderation of very severe solids losses experienced from many package plants is of value, and where long detention times do not create a major land demand. Such ponds do not provide a suitable means of reliably achieving effluent BOD and suspended solids concentrations of less than 20 mg/l and thus cannot be classed as providing an effluent quality any better than that which could be provided by a well-operated secondary plant. The chief virtue lies in the ability to moderate the heavy solids losses from small, poorly operated package plants.

The large space requirements are themselves a strong negative factor when considering application to large plants. Also, there should be no settleable solids losses of the magnitude experienced in small plants. Effluent chlorination can provide more efficient disinfection of the effluent, which negates the only remaining virtue of the polishing pond for larger plants. Regardless of plant size, an effluent quality requirement beyond secondary quality eliminates settling or polishing ponds as a realistic alternate solution.

References

1. Anonymous, "Renovation and Reuse of Waste Waters in Britain." Published by Water Pollution Research Laboratory, 1967.

2. Bologna, A. E., "Solution to a Solids Carryover Problem," *Water Pollution Control Association of Pennsylvania Magazine,* Sept.–Oct., 1969, p. 14.
3. Convery, J. J., "Phosphorus Removal by Tertiary Treatment with Lime and Alum," FWPCA Symposium on Nutrient Removal and Advanced Waste Treatment, Tampa, Florida (Nov. 15, 1968).
4. Culp, G. L., Hansen, S. P., and Richardson, G. H., "High Rate Sedimentation in Water Treatment Works," *Journal American Water Works Assoc.,* 1968, p. 681.
5. Culp, G. L., Hsiung, K. Y., and Conley, W. R., "Tube Clarification Process, Operating Experience," *Journal of the Sanitary Engineering Division,* American Society of Civil Engineers, Oct., 1969, p. 829.
6. Hansen, S. P., and Culp, G. L., "Applying Shallow Depth Sedimentation Theory," *Journal American Water Works Assoc.,* 1967, p. 1134.
7. ———, and Stukenberg, J. R., "Practical Application of Idealized Sedimentation Theory," *Journal Water Pollution Control Federation,* 1969, p. 1421.
8. Hazen, A., "On Sedimentation," *Transactions, American Society of Civil Engineers,* 1904, p. 45.
9. Kalinske, A. A., and Shell, G. L., "Phosphate Removal from Waste Effluent and Raw Wastes Using Chemical Treatment." Presented at the Phosphorous Removal Conference sponsored by the FWPCA, Chicago, Illinois (1968).
10. Livingston, A. P., "High-rate Clarification and Filtration Augment Buffalo Pond Plant Capacity," *Water and Sewage Works,* Apr., 1969, p. 119.
11. McMichael, F. C., "Sedimentation in Inclined Tubes and Its Application for the Design of High-Rate Sedimentation Devices." Presented at the International Union of Theoretical and Applied Mechanics Symposium on Flow of Fluid-Solid Mixtures, University of Cambridge, England (Mar. 24–29, 1969).
12. Morris, G. L., VanDenBerg, L., Culp, G. L., Geckler, J. R., and Porges, R., "Extended Aeration Plants and Intermittent Watercourses," Public Health Service Publication No. 999–WP–8 (July, 1963).
13. Oakley, H. R., and Cripps, T., "British Practice in the Tertiary Treatment of Sewage," *Journal Water Pollution Control Federation,* 1969, p. 36.
14. O'Farrell, T. P., Bishop, D. F., and Bennett, S. M., "Advanced Waste Treatment at Washington, D.C." Presented at the Sixty-fifth Annual AIChE Meeting, Cleveland, Ohio (May, 1969).
15. Rose, J. L., "Removal of Phosphorous by Alum." Presented at the Phosphorous Removal Conference sponsored by the FWPCA, Chicago, Illinois (1968).
16. "Sewage Treatment Plant Design," American Society of Civil Engineers Manual of Engineering Practice No. 36 (1959).
17. Weber, W. J., Jr., Hopkins, C. 3., and Bloom, R., Jr., "Physiochemical Treatment of Wastewater," *Journal Water Pollution Control Federation,* 1970, p. 83.
18. Yao, K. M., "Theoretical Study of High Rate Sedimentation," *Journal Water Pollution Control Federation,* 1970, p. 218.

4

Ammonia Stripping

GENERAL CONSIDERATIONS

Presently, the three principal methods for removal of nitrogen from wastewater are ammonia stripping, selective ion exchange, and microbial denitrification. Of these, ammonia stripping is by far the cheapest, simplest, and easiest to control. Selective ion exchange is a reliable method, but a fairly complex one, and, at the present stage of development, appears to be rather costly. Microbial denitrification is inexpensive, but the process is difficult to sustain and control. For additional means of nitrogen removal see Chapter 10.

However, ammonia stripping has two serious limitations which the designer must recognize: (1) the practical inability to operate the process at ambient air temperatures below 32°F and (2) the deposition of calcium carbonate scale from the water onto the stripping tower fill, which results in loss of efficiency from reduced air circulation and droplet formation and may eventually completely plug the tower. In warm climates, the temperature limitation does not apply. In some locations in cold climates, such as on a flowing stream or river, nitrogen removal may not be required during the winter season. The other problem—that of scaling—appears to be susceptible to control or elimination, and possible methods for doing this will be discussed later.

Under the right climatic conditions and with the proper precautions regarding scale prevention or removal, ammonia stripping is a practical, reliable method for nitrogen reduction. In cold climates, if nitrogen removal is required during freezing weather, then ammonia stripping may be used in warm weather and a supplemental method such as ion exchange provided for use in other seasons. The ammonia concentration in the off gases from a stripping tower even before dispersion in the surrounding air will not exceed about 10 mg/m^3, while the threshold for odor is about 35 mg/m^3. Under most circumstances, there would be no air or water pollution from the off gases. If necessary, ammonia can be recovered from the off gas in a scrubber or by bubbling it through a thin layer of dilute sulfuric acid.

Under unusual conditions where there is a high concentration of sulfur dioxide in the air due to nearby sources of air pollution, ammonia may react with the sulfur dioxide to form an aerosol or fog. Also in large installations, there may be enough water vapor leaving the tower which, if blown down by an unfavorable wind onto road surfaces or other structures in cold weather might form a hazardous layer of ice.

THE AMMONIA STRIPPING PROCESS

In wastewater, either ammonium ions, NH_4^+, or dissolved ammonia gas, NH_3, or both, may be present. At $pH = 7$ only ammonium ions in true solution are present. At $pH = 12$ only dissolved ammonia gas is present, and this gas can be liberated from wastewater under proper conditions. The equilibrium is represented by the equation $NH_4 \rightleftarrows NH_3\uparrow + H^+$. As the *p*H is increased above 7.0, the reaction proceeds to the right.

Figure 4-1 illustrates this relationship at various temperatures. Two major factors affect the rate of transfer of ammonia gas from water to the atmosphere: (1) surface tension at the air-water interface, and (2) difference in concentration of ammonia in the water and the air. Surface tension is at a minimum in water droplets when the surface film is being formed, and ammonia release is greatest at this instant. Little additional gas transfer takes place once a water droplet is completely formed. Therefore, repeated droplet formation and coalescing of the water assists ammonia stripping. To minimize ammonia concentration in the ambient air, rapid circulation of air is beneficial. Air agitation of the droplets may also speed up ammonia release. The ammonia stripping process, then, consists of (1) raising the *p*H of the

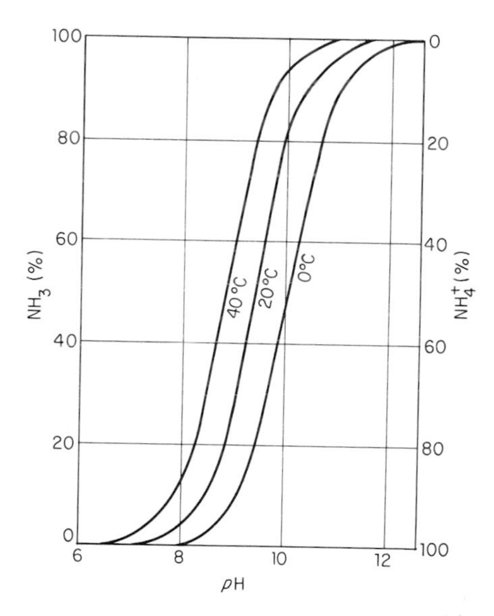

Figure 4–1 Effects of *p*H and temperature on the distribution of ammonia and ammonium ion in water.

water to values in the range of 10.8–11.5, (2) formation and re-formation of water droplets in a stripping tower, and (3) providing air-water contact and droplet agitation by circulation of large quantities of air through the tower.

Ordinarily more than 90 percent of the nitrogen in raw domestic wastewater is in the form of ammonia or compounds from which ammonia is readily formed. If proper environmental conditions are provided in the secondary treatment process, the ammonia will be converted to nitrate. However, this nitrification process can be eliminated by maintaining a relatively high organic loading on the secondary process. It is cheaper and easier to maintain the nitrogen in the form of ammonia than it is to convert it to nitrate. Thus, the removal of nitrogen in the form of ammonia offers some economic and operational advantages over the removal of nitrate-nitrogen.

Kuhn (1956) studied removal of nitrogen from municipal wastewater by ammonia stripping. He found that ammonia-nitrogen removal increased as the *p*H of the wastewater was increased from 8 to 11. No significant increase in the removal of ammonia-nitrogen was obtained by increasing the *p*H from 11–12. Ammonia-nitrogen removals were found to increase as the air/liquid loading was increased from 40–447 cu ft of air per gallon of wastewater. A tower packed with ½-in. Raschig rings and loaded at a rate of 0.3 gpm/ft^2 was used in these studies. The optimum ammonia-nitrogen removal of 92 percent was obtained at a *p*H of 11 with an air/liquid loading of about 500 cu ft/gal and a tower depth of 7 ft.

Prather (1959, 1962, 1964) has reported several studies in which ammonia was removed from petroleum refinery wastewater by stripping with air. The concentrations of ammonia-nitrogen in the untreated wastewater averaged slightly more than 100 mg/l. When 300 cu ft of air were applied per gallon of wastewater, the ammonia removal was found to be 85 percent at a *p*H of 10.5 and 34 percent at a *p*H of 9.4. In another study, in which the wastewater was passed through a closely packed aeration tower with 480 cu ft of air supplied per gallon, ammonia-nitrogen removal by air stripping was found to be very effective (more than 95 percent removal) at any *p*H above 9.0 When the *p*H fell below 9.0, the ammonia-nitrogen removal decreased sharply. The removal fell to 91 percent at a *p*H of 8.9 and to 58 percent at a *p*H of 8.8.

Slechta and Culp (1967) report on the use of a small air conditioning tower designed for home use (Marley Aquatower Model 4411), packed with redwood slats, as a pilot stripping tower. The packing of this tower was compact, having a depth of 2 ft and an area of 5 sq ft. The tower was equipped with a fan which drew 1,520 cfm of air through the tower. This type of tower was used to evaluate the effect of aeration following *p*H adjustment, recognizing that this type of tower packing did not provide optimum air-water contact.

In each of several runs, the *p*H of 22 gal of chemically coagulated and filtered secondary effluent was adjusted with sodium hydroxide. This effluent was then recycled through the tower with frequent measurements of the ammonia-nitrogen content of the tower effluent being made. The *p*H of the wastewater (tower influent) was adjusted to values ranging from 8.0–10.8, and the flow to the tower was varied from 0.5–4 gpm (0.1–0.8 gpm/ft^2).

The efficiency of ammonia-nitrogen removal in the particular aeration tower used in these studies at various *p*H values is shown on Figure 4-2. Optimum ammonia removals at *p*H values above 9.0 were obtained at an aeration time of about 0.5 minute, or, with this particular tower, at an air/liquid loading of about 750 cu ft/gal. With the tower used in this study, the optimum aeration time corresponded to a tower depth of about 20 ft.

Approximately 94 percent of the nitrogen in the applied water was in the form of ammonia-nitrogen. Thus, the 98 percent ammonia-nitrogen removal obtained at a *p*H of 10.8 corresponds to a removal

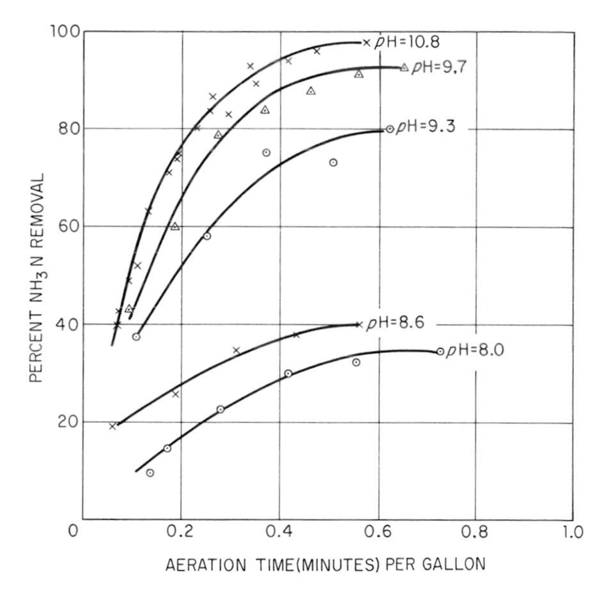

Figure 4–2 Percent ammonia removal versus aeration time. (*From Slechta and Culp, 1967*)

of about 92 percent of all nitrogen present. At a *p*H of 9.7, 93 percent of the ammonia-nitrogen, or 87 percent of all the nitrogen, was removed. Organic nitrogen was not affected by passage through the tower.

Table 4-1 Ammonia-Nitrogen Content of Aeration Tower Influent and Effluent.

TOWER INFLUENT			TOWER EFFLUENT			*Percent NH_3-N Removal*
pH	*Temp. (°C)*	*NH_3-N (mg/l)*	*pH*	*Temp. (°C)*	*NH_3-N (mg/l)*	
8.0	22	28.8	8.0	16	18.2	37
8.6	21	26.6	8.1	17	15.8	41
9.3	23	26.2	8.6	18	5.3	80
9.7	21	30.0	9.1	16	2.1	93
10.8	22	25.5	10.1	17	0.6	98

Table 4-1 summarizes the ammonia-nitrogen content of the aeration tower influent and effluent at various *p*H values. The values shown for tower effluent in Table 4-1 are those occurring at the end of each run shown on Figure 4-2.

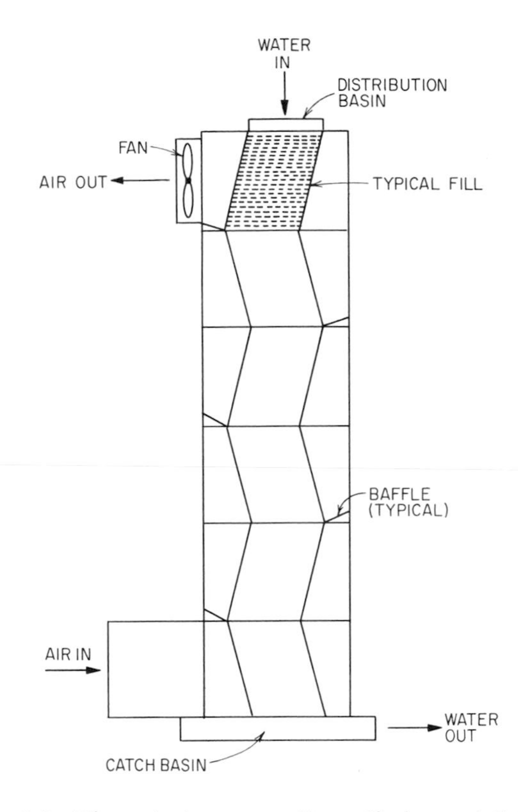

Figure 4–3 Pilot stripping tower. (*From Slechta and Culp, 1967*)

At the conclusion of these initial studies at Tahoe, Smith and Chapman reported (1967) that it was decided that a larger pilot stripping tower was essential to further refine the design criteria for a plant-scale tower. In cooperation with the Marley Company of Kansas City, Missouri, a pilot-scale tower with a larger capacity was designed and constructed.

Figure 4-3 shows a schematic of the ammonia stripping tower. The pilot tower is 25 ft high, 6 ft wide, and 4 ft in depth, and provides countercurrent flow of air. The tower was completely encased except for the air inlet at the bottom and the exit through the fan near the top. The tower contained six sections of close fill, having a cross-sectional area of 7.5 sq ft. The fill, which consisted of redwood slats ⅜ in. thick by 1½ in. wide, was spaced 2 in. on center horizontally and 1½ in. on center vertically. The sections of fill were placed one above the other in a diagonal position, and baffles were provided so that air entering the bottom of the tower was directed almost horizontally through each section of fill before leaving the tower through the fan at the top of the tower. The fan was 42 in. in diameter and was driven by a 5-hp motor.

Secondary effluent was used as a source of water for the ammonia stripping tower. The flow to the tower was measured by a water meter and the flow rate controlled by a gate valve. After metering, the *p*H was raised to a value of 11.5, either by injecting caustic soda into the pressure line feeding the tower, or by pretreating the secondary effluent in a pilot plant using lime to $pH = 11.5$, followed by flocculation and settling. The influent was discharged into a distribution basin at the top of the tower, which had an area of 10 sq ft. There were twenty-four ½-in.-diameter holes evenly spaced over the distribution box floor.

Samples for ammonia analyses were obtained from sampling trays located at various depths in the tower. The surface loading rate and air supply were varied to study the effects of these two factors on the efficiency of ammonia removal for different depths of tower packing.

For a given supply of air per gallon of wastewater treated (see Figure 4-4), the efficiency of ammonia stripping increased with in-

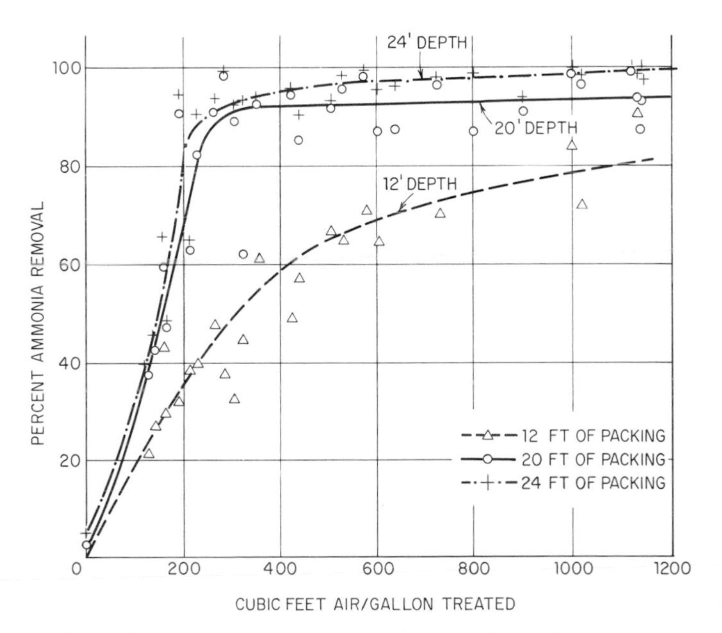

Figure 4–4 Percent ammonia removal versus cubic feet of air per gallon wastewater treated for various depths of packing. (*From Slechta and Culp, 1967*)

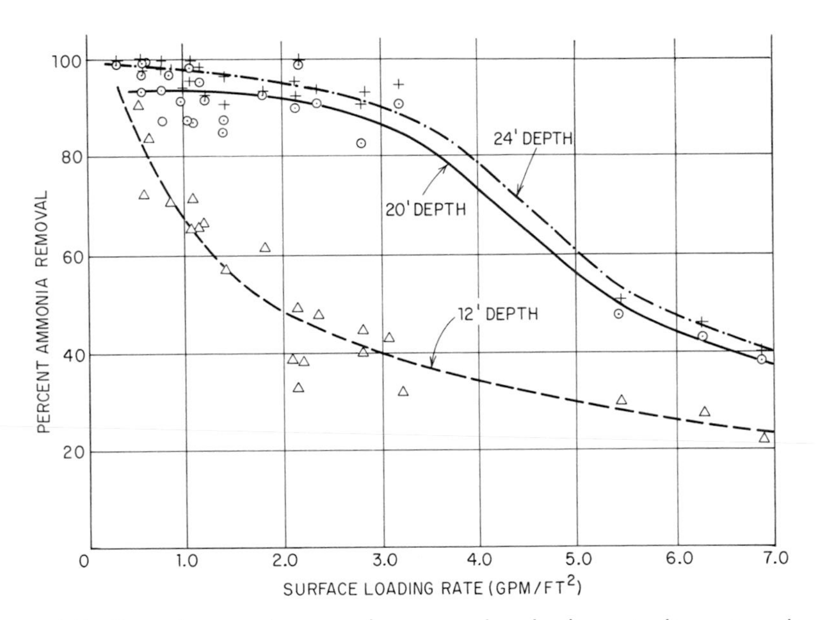

Figure 4–5 Percent ammonia removal versus surface loading rate for various depths of packing. (*From Slechta and Culp, 1967*)

creased packing depths. This appears to be due to the greater number of droplet formations and the resulting increased air-water contact achieved with increased packing depths. The efficiencies of the 20- and 24-ft packing depths were essentially the same for ammonia removals of up to 90 percent. Removals above 90 percent were difficult to achieve with the 20-ft packing, even with increased quantities of air, while it was possible to obtain ammonia removals of 98 percent with the 24-ft packing.

There was little difference between the efficiency of the 20- and 24-ft towers at flow rates up to 7 gpm/ft^2, as shown by Figure 4-5. The rapid decrease in removals for loadings greater than 3 gpm/ft^2 was due to the observed sheeting of water in the tower. Sheeting of the water decreases the number of droplet formations and consequently reduces the efficiency of ammonia stripping.

Figure 4-6 shows that the spacing of the tower fill is very important in determining the air requirements for the ammonia stripping. The 1½ × 2-in. packing has 2⅔ more slats for droplet formation and coalescing than does the 4 × 4 in. packing. The data indicate that droplet formation is a key factor in obtaining high ammonia removal.

The tests showed that the efficiency of ammonia removals de-

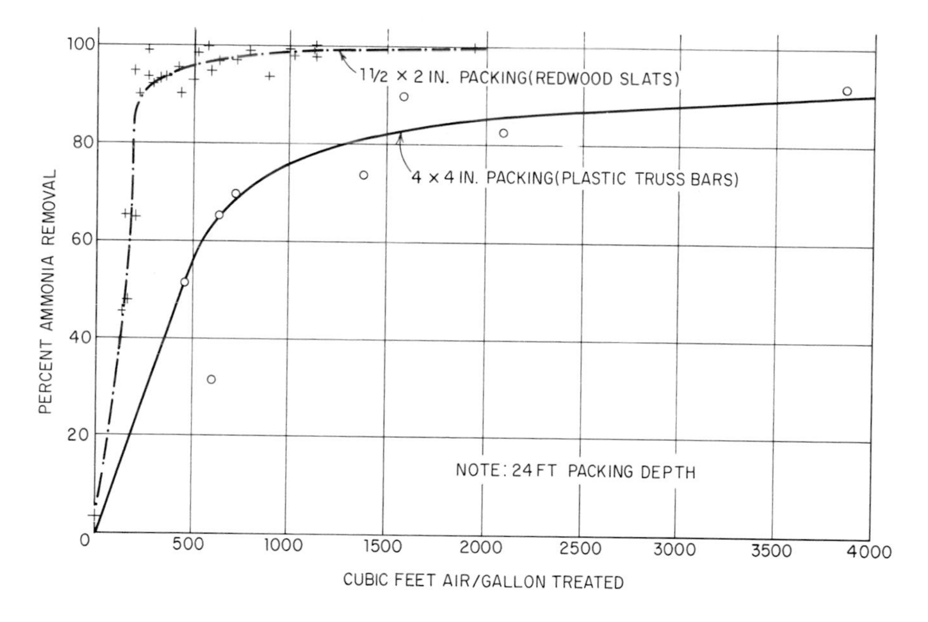

Figure 4–6 Effect of packing spacing on air requirements and efficiency of ammonia stripping. (*From Slechta and Culp, 1967*)

creases more rapidly with increased surface loading rates with the 4 × 4-in. packing than with the 1½ × 2-in. packing. The degree of sheeting was observed to be less in the 1½ × 2-in. packing.

Table 4-2 Air- and Surface-Loading Requirements for 20- and 24-Foot Towers for Various Ammonia-Nitrogen Removals by Ammonia Stripping.

Percent Ammonia-Nitrogen Removal	20-FT DEPTH, 1½" x 2" PACKING		24-FT DEPTH, 1½" x 2" PACKING	
	Ft³ air/gal	*Hydraulic loading gpm/ft²*	*Ft³ air/gal*	*Hydraulic loading gpm/ft²*
80	230	3.6	200	3.9
85	240	3.2	210	3.55
90	280	2.5	250	3.0
95	—	—	400	2.0
98	—	—	800	0.8

Effluent water temperature = 20°C.

Table 4-2 summarizes the air and hydraulic loading requirements for various ammonia-nitrogen removals using 20- and 24-ft towers with a 1½ × 2-in. packing. For the same ammonia-nitrogen removal, it was possible to use less air and higher surface loading rates with the 24-ft tower.

It can be seen that 90 percent ammonia-nitrogen removal was possible with approximately 250 cu ft of air per gallon treated, while the quantities of air required for higher removals increase rapidly (400 cu ft/gal for 95 percent removal and 800 cu ft/gal for 98 percent removal).

The above data and results were obtained during summer operating conditions when the temperature of the wastewater being treated was always above 20°C and ambient air temperatures were high. The tower was later operated during winter conditions in conjunction with a lime clarification unit. At that time, it was observed that the colder air and water temperatures had a pronounced effect on the ammonia stripping efficiency.

Theoretically, the rate of ammonia stripping should be a function of the difference in partial pressure of the ammonia in the liquid phase and the gaseous phase. Ammonia partial pressure data are not available for concentrations as low as those in municipal wastewater. The data available for higher concentrations indicate, however, that a 10°C decrease in process water temperature would decrease the difference in partial pressures by about 40 percent. This decreases the driving force for ammonia stripping by this amount.

Figure 4-7 presents ammonia removal data collected with all oper-

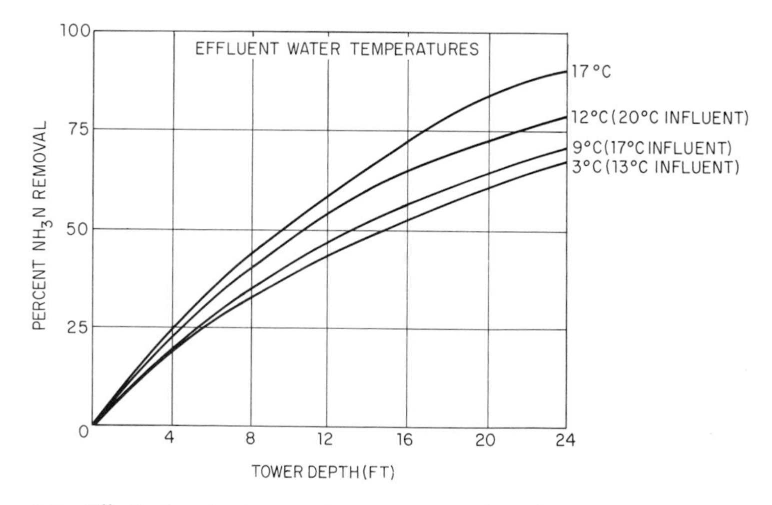

Figure 4–7 Effect of water temperature on ammonia stripping (operating conditions —2.0 gpm/ft², 480 cu ft air/gal). (*From Smith and Chapman, 1967*)

ating conditions except temperature held constant. Also included for comparative purposes, are stripping data collected under similar circumstances during the previous summer. This figure illustrates quite well the decrease in efficiency resulting from a decrease in operating temperature.

The removal efficiency of the tower was substantially reduced as the temperature fell below 20°C, but a lower limit of removal efficiency was apparently reached as the temperature approached 0°C. A change in effluent temperature from 3–9°C improved ammonia removal by only 5 percent, while a change in temperature from 9–12°C resulted in an 8 percent increase in ammonia removal. Data collected from operation of the pilot tower during winter conditions indicate that the average lower limit of the process will be in the range of 50–60 percent ammonia removal.

From these pilot tests, it appeared that a taller tower, or provisions for tower effluent water recirculation, would be beneficial under winter conditions. After 24 ft of packing contact, a limit for ammonia removal had not been reached at any temperature level. The rate of removal diminishes with decreasing temperature and ammonia concentration. Therefore, the effect of providing additional packing contact should be investigated.

In pilot plant tests of ammonia stripping lime-treated wastewater at different locations, the amount and hardness of calcium carbonate scale formation varied widely. At South Tahoe, California, and at Windhoek, South Africa, the scale formed in countercurrent airflow pilot towers was soft and did not accumulate, apparently being washed off continuously by the flow of water. In similar tests at Chicago and Washington, D.C., an extremely hard scale was formed which was very difficult or impossible to remove from the tower packing and structural members. The physical or chemical factors which account for these very important variations in the amount and character of the scale formed are not known at this time.

PLANT-SCALE AMMONIA STRIPPING EXPERIENCE

Upon completion of the second set of pilot plant tests at South Tahoe, it was decided to build a plant-scale unit for purposes of further experimentation and development. The rated plant capacity was 7.5 mgd. However, the original stripping tower as installed was given a nominal rated capacity of only half this amount, or 3.75 mgd. This was done so that any improvements in design or construction

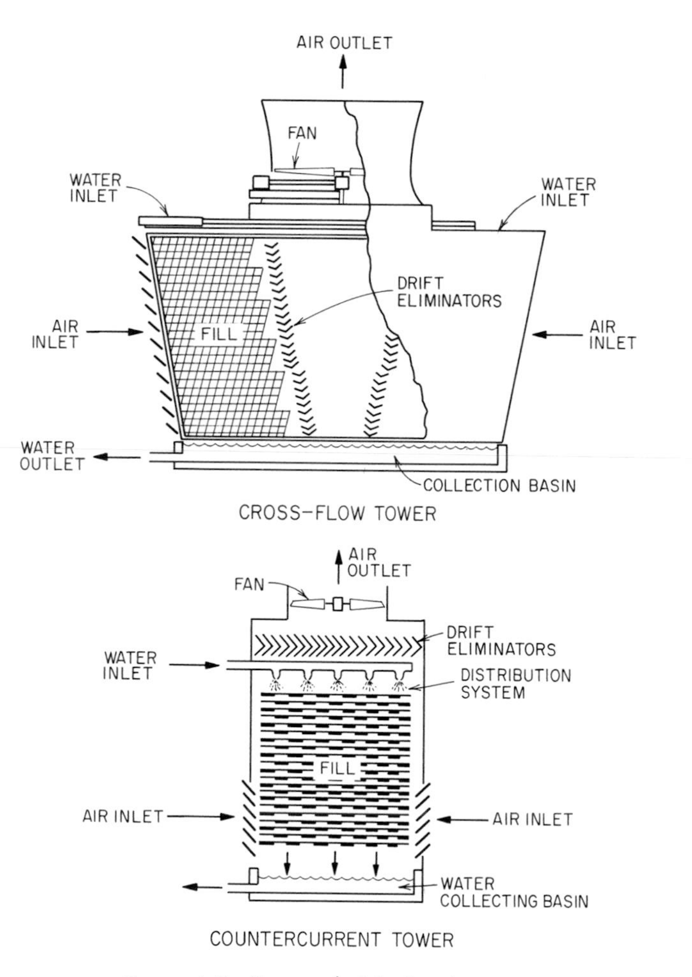

Figure 4–8 Types of stripping towers.

which became evident in operation of the first section of the tower could be utilized later in planning and building the second section. Stage construction has the disadvantage that only part of the total plant flow can be treated in the tower during peak hourly flow rates, and the remainder of the total flow must be bypassed around the ammonia stripping tower at that time. Even though the pilot towers had both been countercurrent airflow towers, the full-scale tower was designed as a cross-flow unit. The general features of these two types of structures are shown in Figure 4-8. A full-scale countercurrent lower was under construction in 1970 at the Pretoria, South Africa, plant. The internals of the Tahoe stripping tower can be seen from Figure 4-9.

Figure 4–9 Ammonia stripping tower under construction at South Tahoe. (*Courtesy Cornell, Howland, Hayes & Merryfield*)

Table 4-3 Design Data for Full Scale Ammonia Stripping Tower.

Capacity: Nominal, 3.75 mgd

Type: Cross flow with central air plenum and vertical air discharge through fan cylinder at top of tower.

Fill: Plan area, 900 sq ft
Height, 24 ft
Splash bars:
material, rough-sawn treated hemlock
size, 3/8 x 1½ in.
spacing, vertical 1.33 in.
horizontal 2 in.

Air Flow: Fan, two-speed, reversible, 24-ft diameter, horizontal

Water Rate		Air Rate	
gpm	gpm/ft²	cfm	cfm/gpm
1,350	1.0	750,000	550
1,800	2.0	700,000	390
2,700	3.0	625,000	230

Tower Structure: Redwood
Tower Enclosure: Corrugated cement asbestos
Air Pressure Drop: ½ in. of water at 1 gpm/ft²

The principal design features of the Tahoe ammonia stripping tower are given in Table 4-3.

The completed tower is shown in the photograph of Figure 4-10. The overall dimensions of the tower are 32 × 64 × 47 ft high. Water at $pH = 11$ is pumped to the top of the tower by either or both of two constant-speed pumps. These pumps are backflushed two or three times daily to minimize buildup of calcium carbonate scale in the pump units. When the plant inflow is less than the rate at which the pumps are delivering water to the tower, some water is recycled from the tower effluent back to the pump suction well. This avoids the need for variable speed pump control, and at the same time provides some recirculation through the tower, which improves ammonia removal. The flow pattern from the chemical clarifier to the tower pumps, through the tower, and through the two-stage recarbonation system, is shown schematically in Figure 4-11.

At the top of the tower, the influent water enters a covered distribution box and overflows to a distribution basin. The distribution basin is a flat deck with a series of holes fitted with plastic nozzles.

Figure 4–10 3.75 mgd ammonia stripping tower at South Tahoe. (*Reprinted from* Water & Wastes Engineering, *6:4. Apr., 1969, p. 36. R. H. Donnelley Corp.*)

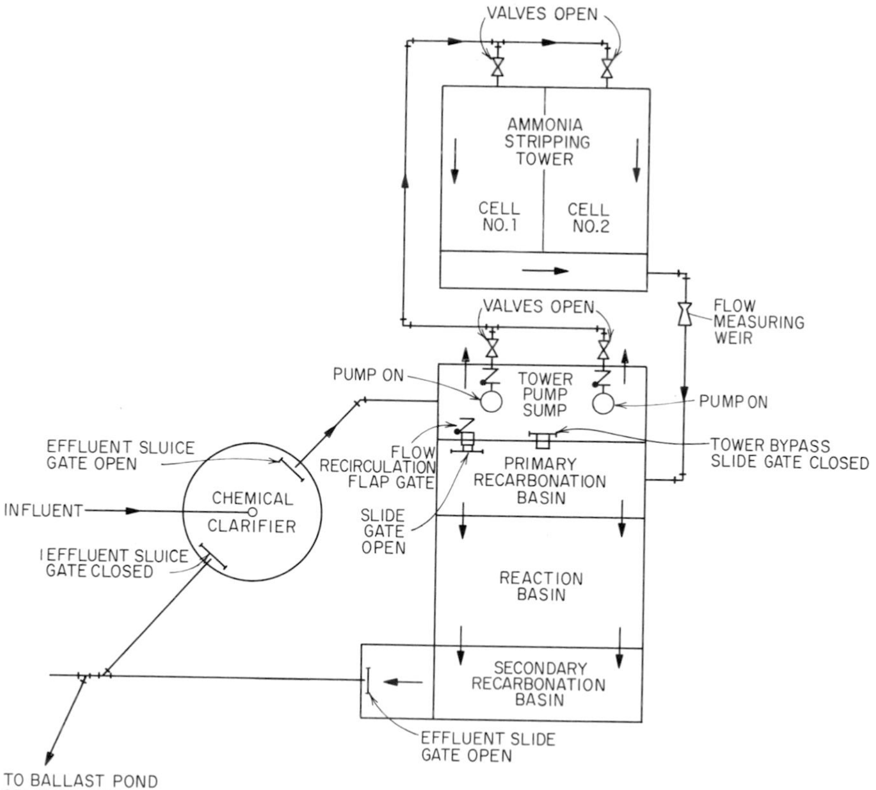

Figure 4–11 Ammonia stripping tower and recarbonation basins. (*Courtesy Clair A. Hill & Assocs.*)

Further distribution of the inflow is provided by diffusion decks immediately below the distribution basin. At 6-ft vertical intervals in the fill there are three other diffusion decks. The tower fill proper provides, theoretically, 215 successive droplet formations as the water passes down through the tower. The tower effluent falls into a concrete collection basin which also forms the base for the tower structure. From the collection basin, the tower effluent passes through a measuring weir into the first-stage recarbonation chamber, where, as previously mentioned, excess pumpage returns through a flap gate into the tower pump sump to be recirculated through the tower.

Air enters the tower through side louvers, passes horizontally through the tower fill and drift eliminators (or airflow equalizers), and enters a central plenum. At the top center of the plenum is a 24-ft diameter, 6-bladed, horizontal fan. Fan blades and fan cylinder are both made of glass-reinforced polyester. The fan takes suction from the plenum and discharges to the atmosphere through the fan cylinder. The fan has a maximum capacity of about 750,000 cfm. It is equipped with a two-speed, reversible 100-hp motor.

One of the greatest advantages of this method of nitrogen removal is its extreme simplicity. Water is merely pumped to the top of the tower at $pH = 11$, air is drawn through the fill, and the ammonia is stripped from the water droplets. The only control required is to maintain the proper pH in the influent water. This simplicity of operation also very much enhances the reliability of the process.

Performance of the full-scale tower closely parallels that of the pilot tower insofar as ammonia removals are concerned. At similar hydraulic loadings, airflows, and water and air temperatures, the results with the full-scale cross-flow tower are much the same as with the pilot countercurrent tower. However, a much greater insight has been gained regarding practical operating problems, and the limitations and advantages of the ammonia stripping process.

The operating problems are freezing and calcium carbonate scale formation. The freezing problem was, of course, recognized in pilot plant operation, but could not be fully evaluated until the full-scale tower was operated at low air temperatures. Cooling towers similar in design to ammonia stripping towers operate very successfully in severe winter climates in the northern United States and Canada. This is achieved in several ways. One method is to use large flow distribution orifices at the outside face of the tower. This concentrates a curtain of warm water at the point where the cold air first enters the tower and gives some protection against freezing. If there is a slight ice buildup on the outside tower face, the draft fan can be reversed, thus blowing warm inside air to the frozen area and melting the ice. These are satisfactory ways to prevent freezing or to correct it in a cooling tower, because the desired cooling of the water is still obtained. These same methods work equally well to prevent freezing in ammonia towers, but there is considerable sacrifice in ammonia removal efficiency. The tower at Tahoe has been operated without serious icing at air temperatures as low as 20°F, but ammonia removals drop to less than 30 percent, which is of questionable benefit. The operation of the plant-scale tower has demonstrated that it is not practical to operate the tower at air temperatures below 32°F. It also is not practical to heat the air or water in order to prevent ice formation on the tower because of the tremendous quantities of heat required.

The second problem, that of calcium carbonate scale deposition on the tower fill and structural members was not anticipated, because it did not occur in the countercurrent pilot tower. The fill in the pilot tower was checked very carefully periodically over several months of operation with lime adjustment of *p*H, and there was no evidence of scale buildup. Similar results have been reported after nine months' operation of a pilot countercurrent tower at Windhoek. There is definitely a problem, however, in the full-scale cross-flow tower at Tahoe, but this problem is not an insurmountable one. The scale which forms is very soft and friable, and can be removed quite readily by light hosing with a water spray, or by rodding through the tower. In ammonia stripping of waters which form this soft scale, it is a simple matter to provide for easy, inexpensive removal of the deposits by providing complete accessibility to the fill, or by installing a system of water sprays specifically for cleaning purposes, to be operated at intervals as required. So actually, by anticipating this soft scale problem in the design of the tower the problem can be overcome at very low cost. Because of the inaccessibility of parts of the fill in the Tahoe tower due to the location of certain structural members, this tower has been cleaned with a mixture of dilute sulfuric acid and an organic dispersant. This is another satisfactory solution to the problem. The cost of chemicals for one cleaning of this tower is less than $200 not including the cost of disposing of the used chemical solution. Since only one tower unit is installed at present, cleaning does involve shut down of ammonia stripping operations until it is completed. For full-time uninterrupted service, at least two units should be provided.

Hard scale deposits such as those formed by the waters at Chicago and Washington, D.C. pose a much more difficult problem, because the scale cannot be removed by water spraying or even by hard scraping.

The capital cost of the ammonia stripping (1969 FWPCA STP Construction Cost Index = 127.1) installation at Tahoe for 3.75 mgd capacity is $310,000, or about $8.00/mg if the investment is amortized at 5 percent interest over 25 years. Costs for operation and maintenance are about $8.75/mg.

References

1. Cillie, G. G., *et al.*, "The Reclamation of Sewage Effluents for Domestic Use," Third International Conference on Water Pollution Research, Munich, Germany, Section II, Paper I, WPCF, Washington, D.C., 1966.

2. Culp, R. L., and Moyer, H. E., "Wastewater Reclamation and Export at South Tahoe," *Civil Engineering,* 39:38 (June, 1969).
3. Kuhn, P. A., "Removal of Ammonium Nitrogen from Sewage Effluent," unpublished MS Thesis, University of Wisconsin, Madison, Wisconsin (1956).
4. Prather, B. V., "Wastewater Aeration May Be Key to More Efficient Removal of Impurities," *Oil and Gas Journal,* Nov. 30, 1959, p. 78.
5. ———, "Chemical Oxidation of Petroleum Refinery Wastes." Presented at the Thirteenth Industrial Wastes Conference, Oklahoma State University, Stillwater, Oklahoma (Nov., 1962).
6. ———, and Gaudy, A. F., "Combined Chemical, Physical, and Biological Processes in Refinery Wastewater Purification." Presented at the Twenty-ninth Midyear Meeting of the American Petroleum Institute's Division of Refining, St. Louis, Missouri (May, 1964).
7. Sawyer, C. N., *Chemistry for Sanitary Engineers,* McGraw-Hill Book Co., New York, 1960.
8. Slechta, A. F., and Culp, G. L., "Water Reclamation Studies At The South Public Utility District," *Journal Water Pollution Control Federation,* 1967, p. 787.
9. Smith, C. E., and Chapman, R. L., "Recovery of Coagulant, Nitrogen Removal, and Carbon Regeneration in Wastewater Reclamation." Final Report by South Tahoe PUD to FWPCA, Demonstration Grant WPD–85 (June, 1967).
10. Snow, R. H., and Wnek, W. J., "Ammonia Stripping Mathematical Model For Wastewater Treatment," Final Report (Report No. IITRI–C6152–6), of IIT Research Institute to FWPCA (Dec., 1968).

5

Recarbonation

PURPOSE

Recarbonation is a unit water treatment process which has been in use for many years in numerous municipal and industrial lime-soda softening plants throughout the world. More recently, with lime treatment of wastewaters, particularly with the massive doses of lime used in phosphorus removal, recarbonation has seen increasing use in wastewater treatment and water reclamation plants.

The addition of sufficient lime to wastewater raises the *p*H and converts bicarbonates and carbonates to hydroxides, incidental to the main purposes of treatment which are coagulation of nonsettleable matter and removal of phosphorus. Recarbonation is a term applied to the addition of carbon dioxide to a lime-treated water. When carbon dioxide is added to high *p*H, lime-treated water, the *p*H is lowered and the hydroxides are reconverted to carbonates and bicarbonates. Thus, the term recarbonation is very descriptive of the result of adding carbon dioxide to wastewater.

The basic purpose of recarbonation is the downward adjustment of the *p*H of the water. Properly done, this places the water in calcium carbonate equilibrium, and avoids problems of deposition of calcium scale which would occur without the reduction in *p*H accomplished by recarbonation. In water works practice, the carbon dioxide is added to the water ahead of the filters in order to avoid coating of the grains of filter media with calcium carbonate, which would eventually increase the grain size to the point that filter efficiency would be reduced. In water works it is also important to lower the *p*H of the lime-treated water to the point of calcium carbonate stability to avoid deposition of calcium carbonate in pipelines.

In advanced treatment of wastewaters, *p*H control by recarbonation has even greater significance than in water works practice, because of the *p*H effects on treatment processes commonly found downstream.

There are several reasons for adjusting the *p*H of wastewater during treatment. Coagulation, flocculation, and clarification of wastewaters can be accomplished in a rather wide range of *p*H values but particular coagulants or coagulant aids ordinarily yield optimum results within a rather narrow range of *p*H values. Lime usually gives good results in domestic wastewaters at *p*H values above 9.6 or 9.8. Alum is ordinarily good at *p*H values below 7.3 or above 8.6 The effective *p*H range for coagulants usually can be broadened by use of the correct polymer, activated silica, or other coagulant aids. Ammonia stripping is best at *p*H values above 10.8, because then most of the nitrogen is in the form of dissolved ammonia gas rather than as ammonium ions in true solution. Generally speaking, filtration through granular media is best in the *p*H range of 6.5–7.5, but the filterability of water depends upon many other factors, including the physical and chemical characteristics of the water, the coagulant used, and the chemicals employed as filter aids, if any. Activated granular carbon adsorption of organics from water generally takes place in the *p*H range of 5–9, with better adsorption at values below 7. Very often, if the *p*H is above 9, desorption will occur. That is, organics which have been adsorbed previously on the carbon will be released to the high *p*H water. For example, a column of granular activated carbon which has been operating at a *p*H of 7 for several days with good removal of color from the applied water may still have capacity for further color removal at a *p*H of 7. However, if water with a *p*H of 10 is passed through the column, the color of the effluent water from the column will undoubtedly be higher than that of the influent, due to desorption.

Massive lime treatment of wastewaters for phosphorus removal often raises the *p*H to values of about 11. Primary recarbonation is used to reduce the *p*H from 11–9.3, which is near that of minimum solubility for calcium carbonate. In domestic wastewater, primary recarbonation to *p*H = 9.3 results in the formation of a heavy, rapidly settling floc which is principally calcium carbonate, although some phosphorus is also removed from solution by adsorption on the floc. If sufficient reaction time, usually about 15 min in cold water, is allowed for the primary recarbonation reaction to go to completion, the calcium carbonate floc does not redissolve with subsequent further lowering of *p*H in secondary recarbonation. (However, there is a tendency for the magnesium salts to do so.) If lime is to be reclaimed by recalcining and reused, this settled primary recarbonation floc is a rich source of calcium oxide, and may represent as much as one third of the total recoverable lime. The second stage of recarbonation to *p*H = 7, is beneficial in several ways: it prepares the water for filtration; it lowers the *p*H to a value which increases the efficiency of carbon adsorption of organics, to an excellent range for effective disinfection by chlorine, and to a value suitable for discharge; and it stabilizes the water in respect to scale formation in pipelines. If the *p*H were not reduced to less than about 8.8 before application to the filters and carbon beds, extensive deposition of calcium carbonate would occur on the surface of the grains. This could reduce filter efficiency, and could also drastically reduce the absorptive capacity of granular activated carbon for organics. It would produce rapid ash buildup in the carbon pores upon regeneration of the carbon, and would lead to early replacement of the carbon.

SINGLE-STAGE VERSUS TWO-STAGE RECARBONATION

It is possible to reduce the *p*H of a treated wastewater from 11–7 or to any other desired value in one stage of recarbonation. Single-stage recarbonation eliminates the need for the intermediate settling basin which is used with two-stage systems. However, by applying sufficient carbon dioxide in one step for the total *p*H reduction, little, if any, calcium is precipitated with the bulk of calcium remaining in solution, thus increasing the calcium hardness of the finished water, and, in addition, causing the loss of a large quantity of calcium carbonate which could otherwise be settled out, recalcined to lime, and reused. If lime is to be reclaimed or if calcium reduction in the effluent is desired, then two-stage recarbonation is required. Otherwise, single-stage recarbonation may be used with some savings in initial cost, and some reduction in the amount of lime sludge to be handled.

SOURCES OF CARBON DIOXIDE

In advanced wastewater treatment plants, the usual source of carbon dioxide for recarbonation probably will be the stack gas from either a lime recalcining furnace or a sludge incineration furnace. Other possible sources include the use of commercial liquid carbon dioxide; or the burning of natural gas, propane, butane, kerosene, fuel oil, or coke.

The stack gas from a sludge incineration furnace which is fired with natural gas will contain about 16 percent carbon dioxide on a wet basis or about 10 percent on a dry basis. About 10 percent CO_2 on a dry basis is usually used for design purposes. The burning of 1,000 cubic feet of natural gas produces about 115 lb of CO_2. Artificial gas produces about 80 lb of CO_2 per 1,000 cu ft burned.

The stack gas from a lime recalcining furnace contains not only the carbon dioxide (CO_2) produced by combustion of the fuel, but, in addition, the CO_2 driven off of the calcium carbonate sludge in the recalcining process. For design purposes, a value of 16 percent CO_2 in lime furnace stack gas is a conservative figure to use.

Kerosene and No. 2 fuel oil will yield about 20 lb of CO_2 per pound of fuel. Coke will produce approximately 3 lb of CO_2 per pound of coke burned. Commercial liquid CO_2 contains 99.5 percent of CO_2.

QUANTITIES OF CARBON DIOXIDE REQUIRED

In recarbonation, one molecule of CO_2 is required to convert calcium hydroxide (caustic alkalinity) to calcium carbonate. In addition, it takes one molecule of CO_2 to convert calcium carbonate to calcium bicarbonate. It follows then, that two molecules of CO_2 are required to convert calcium hydroxide to calcium bicarbonate. These reactions are represented by the following equations:

$$Ca(OH)_2 + CO_2 \rightarrow CaCO_3 + H_2O$$
$$CaCO_3 + CO_2 + H_2O \rightarrow Ca(HCO_3)_2$$

Since all forms of alkalinity are expressed in terms of calcium carbonate (molecular weight = 100), the calculations are as follows:

1. Calcium hydroxide to calcium carbonate:
   (molecular weight of CO_2 = 44, and 1 mg/l = 8.33 lb/mg)
   CO_2 (in lb/mg)
   $$= \frac{44}{100} \times 8.33 \times (OH^- \text{ Alk. in mg/l as } CaCO_3)$$
   $$= 3.7 \times (OH^- \text{ Alk. in mg/l as } CaCO_3)$$

2. Calcium carbonate to calcium bicarbonate:
   CO_2 (lb/mg)
   $= \frac{44}{100} \times 8.33 \times (CO^{=}_3$ Alk. in mg/l as $CaCO_3)$
   $= 3.7 \times (CO^{=}_3$ Alk. in mg/l as $CaCO_3)$
3. Then, for calcium hydroxide to calcium bicarbonate:
   CO_2 (lb/mg)
   $= 7.4 \times (OH^-$ Alk. in mg/l as $CaCO_3)$

Sample Calculations for the Amount of CO_2 Required

Assume that 400 mg/l of calcium oxide (CaO) have been added to a sample of a typical domestic wastewater and that the stirred and decanted liquor has the following characteristics:

*p*H = 11.7
Alkalinities (in mg/l as $CaCO_3$):
  OH^- = 380
  $CO^{=}_3$ = 120
  HCO^-_3 = 0

After first recarbonation in the laboratory using bottled carbon dioxide, analysis of the lime-treated wastewater shows the following:

*p*H = 9.3
Alkalinities (in mg/l as $CaCO_3$):
  OH^- = 0
  $CO^{=}_3$ = 180
  HCO^-_3 = 380

Then, after secondary recarbonation in the laboratory, the water has the following analysis:

*p*H = 8.3
Alkalinities (in mg/l as $CaCO_3$):
  OH^- = 0
  $CO^{=}_3$ = 0
  HCO^-_3 = 750

To change all caustic alkalinity and all carbonate alkalinity to bicarbonates then, the amount of CO_2 required is as calculated below:

$7.4 \times 380 = 2{,}812$ lb/mg of CO_2
$3.7 \times 120 = 444$
Total = 3,256 lb/mg of CO_2

For a l mgd flow, 3,256 lb of CO_2 per day are required. If it is assumed that the CO_2 content of the stack gas to be used for recarbonation is 10 percent, then $\frac{100}{10} \times 3,256 = 32,560$ lb of stack gas must be compressed in order to supply the necessary CO_2 to recarbonate 1 million gal of wastewater.

At standard conditions of 14.7 psia and 60°F (520° absolute), assume that the density of the stack gas is the same as for CO_2, or 0.116 lb/ft^3. Then, $\frac{32,560 \text{ lb}}{0.116 \text{ lb/ft}^3} = 280,600$ cu ft of stack gas are required to recarbonate 1 mg of wastewater, at standard conditions. This is equal to 195 cfm/mgd. These figures at standard conditions must be adjusted for temperature and altitude.

If it is assumed that the stack gas is cooled in scrubbing to 110°F, then the temperature correction is $\frac{110 + 460}{60 + 460} \times 195 = 1.1 \times 195 = 215$ cfm/mgd at sea level.

With the same gas temperature, for a plant at 6,300 ft above sea level (11.6 psia), the altitude correction is $\frac{14.7}{11.6} \times 215 = 1.26 \times 215 = 270$ cfm/mgd.

Since some CO_2 is not absorbed in the water but escapes at the water surface, it is customary to add about 20 percent to the theoretical requirements. If this is done then, at sea level 260 cfm/mgd of blower or compressor capacity is required, and at 6,300 ft of altitude 325 cfm/mgd of plant capacity is needed.

It must be emphasized that these calculations are based on the following assumptions:

1. The water to be recarbonated has a *p*H = 11.7 with OH^- Alk. = 380 mg/l as $CaCO_3$, and with $CO^=_3$ alkalinity = 120 mg/l as $CaCO_3$, all of which is to be converted to bicarbonates.
2. There is 10 percent CO_2 in the gas.
3. The flue gas temperature is 110°F.
4. An excess of 20 percent is added to the theoretical values to allow for absorption losses.

Conditions at each installation undoubtedly will differ from the assumptions used here, and calculations must be based on actual values rather than those given.

It should be noted that the quantities of CO_2 required for this type of wastewater treatment are many times greater than the amounts used in water softening plants. However, the observed actual CO_2 consumption at the South Tahoe Public Utility District Water Reclamation Plant corresponds quite closely to the figures obtained using the above method of computation.

NONPRESSURIZED CARBON DIOXIDE GENERATORS

If stack gas from a furnace operating at atmospheric pressure is to be used as a source of CO_2, the gas should be passed through a wet scrubber. Wet scrubbers provide contact between the gas and a flow of scrubbing water. Particulate matter is removed from the gas, and the gas is cooled. Wet scrubbers may be one of three general types; impingement, Venturi, and surface area. Figure 5–1 shows two water jet impingement type scrubbers as manufactured by W. W. Sly Manufacturing Company, Cleveland, Ohio, and as installed at the South Tahoe Public Utility District Plant. One serves a sludge incineration furnace (right), and the other a lime recalcining furnace. These scrubbers are very efficient in removing potential air pollutants from the exhaust gas, and provide some protection of the CO_2 compression equipment against plugging or scaling by particulates. The scrubbers cool the hot stack gas down to about 110°F.

Figure 5–1 Water jet impingement type stack gas scrubbers installed on sludge incineration and lime recalcining furnaces. *(Courtesy Cornell, Howland, Hayes, & Merryfield)*

When stack gas is used as the source of CO_2, the stack gas supply must exceed the maximum demands for CO_2. With this situation, control of the amount of CO_2 applied to the water is very simple. Air may be admitted through a valve into the suction line leading to the compressor as required to reduce the amount of CO_2 to that desired, or, as an alternate, part of the compressed gas may be bled off to the atmosphere through a valve in the compressor discharge line. As another method of control, compressed gas may be recirculated from the compressor discharge line back to the suction line through a bypass line and control valve. However, this method has the serious disadvantage of warming the gas due to the heat of compression, and excessive recirculation can lead to compressor damage by overheating or increased corrosion at the elevated temperatures.

COMPRESSOR SELECTION

Even with thorough scrubbing, stack gas from sludge incineration furnaces or lime recalcining furnaces will contain sufficient particulate matter to cause plugging and seizure problems in some types of blowers and compressors, particularly those with limited clearance between moving metal parts. This problem is less severe when using stack gas from atmospheric furnaces, which burn fuel primarily for production of CO_2.

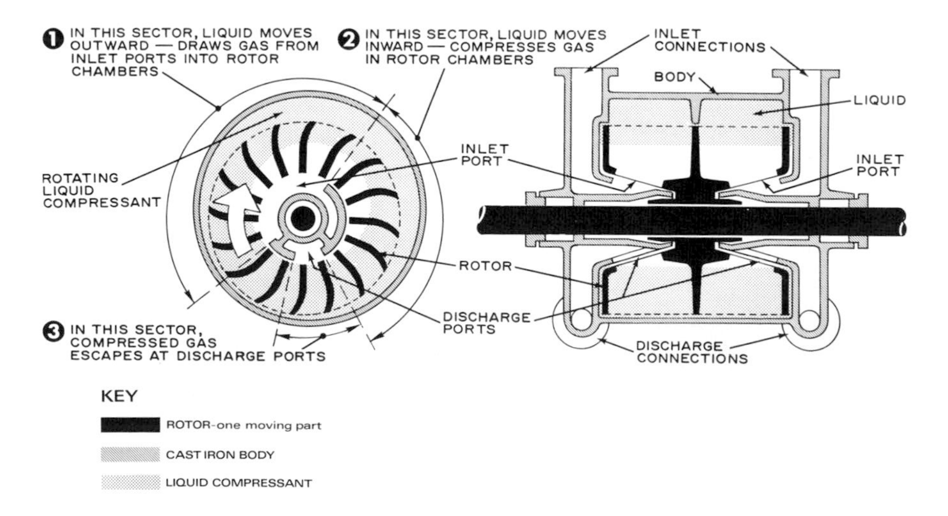

Figure 5–2 Section through water-sealed carbon dioxide compressor. *(Courtesy Nash Engineering Co.)*

Water-sealed compressors similar to wet vacuum pumps are a good selection for handling dirty, corrosive gases. This type of compressor consists of a squirrel cage type of rotor which revolves in a circular casing containing water. The rotor shaft is located off-center toward the bottom of the casing. Figure 5–2 is a cross-section through a water-sealed compressor. As the rotor revolves, centrifugal force pushes the water out against the sides of the casing, leaving a dough-nut-like hole of air in the center. Because the center drum of the rotor is located eccentrically, there is a large air space above it and none below. Thus, as air is driven from the large space into a progressively smaller one, it is compressed. Figure 5–3 shows an exterior view of a water-sealed CO_2 compressor and disassembled parts. This is a simple, reliable piece of equipment with only one moving part. It has increased capacity when handling hot, saturated vapors, since the vapors are condensed by the cool liquid compressant. The water-sealed compressor is a relatively quiet-running unit, free from pulsations and vibrations.

If the CO_2 distribution grids are submerged a minimum of 8 ft in water, as they usually are, the CO_2 compressor must deliver against a differential pressure across the machine of about 6–8 psi. The exact rating must be determined by calculation, taking into account not only the depth of submergence of the distribution piping, but also orifice losses and pipe friction losses. This is discussed in more detail later, because it is common to all types of CO_2 systems. The compressors may be of cast-iron construction, or may be supplied with bronze rotor and cones at considerable extra cost. The following accessories are commonly required with water-sealed compressor units: water separator with gage glass and bronze float valve; discharge check valve; expansion joints for inlet and outlet piping; water seal supply line with adjusting cock and orifice union, water line strainer, inlet water spray nozzles, and seal-water line solenoid valve. In addition, the discharge line is usually fitted with an automatic pressure relief valve and a bleed-off valve, both of which should discharge to free atmosphere. It is not good practice to install shutoff or isolation valves on either the compressor suction or discharge lines because of the possibility of serious damage to compressor or pipelines in the event that the compressor is operated in error with either or both of such valves closed.

In selecting CO_2 compressor units to meet total capacity requirements, it is a good idea, except in very small installations, to provide at least three compressor units. By properly sizing these then it is possible to satisfy two needs—to secure a range in output and to

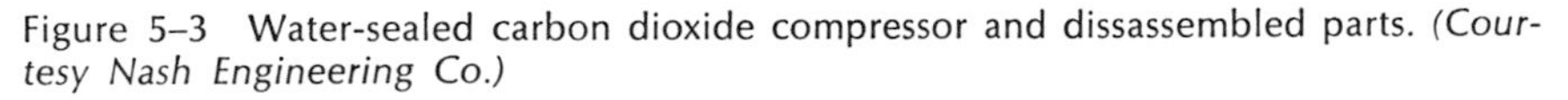

Figure 5–3 Water-sealed carbon dioxide compressor and dissassembled parts. *(Courtesy Nash Engineering Co.)*

provide standby service. If it is assumed that the total CO_2 capacity required is 1,500 cfm, then units with individual capacities of 500, 1,000, and 1,500 cfm would represent a good choice. This combination gives a range of 500 to 1,500 cfm to match plant needs, and it supplies complete standby with the largest unit out of service by using the two smaller units together. A typical recarbonation system using stack gas at atmospheric pressure is illustrated by Figure 5–4. As indicated in this figure, automatic *p*H control of the recarbonated

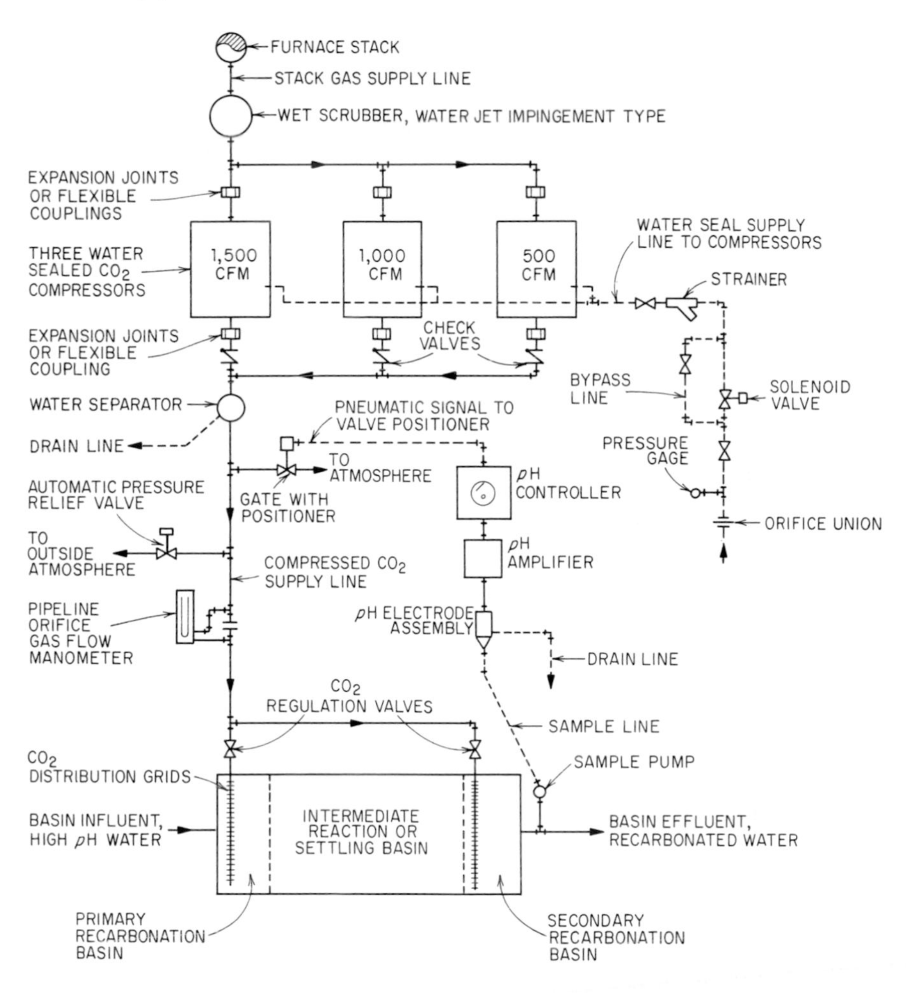

Figure 5–4 Typical recarbonation system using stack gas.

effluent can be provided by continuously monitoring an effluent sample for *p*H to operate a *p*H controller, which in turn positions a bleed-off valve in the CO_2 compressor discharge line to limit the amount of CO_2 to that necessary to maintain the desired *p*H.

PRESSURE GENERATORS AND UNDERWATER BURNERS

Generators designed specifically for the production of carbon dioxide for recarbonation are usually either pressure generators or submerged underwater burners. Most early installations were of the atmospheric type in which the fuel is burned at atmospheric pressure and the off-gas is scrubbed and compressed. These systems are expensive to maintain because of the corrosive effects of the hot, moist combustion gases, and atmospheric generators have largely been replaced by pressure generators and underwater burners, except where waste stack gas is available from another source. Both types of pressure CO_2 generation equipment are now in commercial manufacture.

Pressure or forced-draft generators produce CO_2 by burning natural gas, fuel oil, or other fuels in a pressure chamber. The fuel and excess air are first compressed and injected, and then burned at a pressure which is sufficiently high to allow discharge directly into the water to be recarbonated. The compressors handle only dry gas or dry air at ambient temperatures and thus the corrosion problems involved in handling the hot, moist stack gases are avoided. One difficulty with this type of pressure generator is its limited capacity range, which may be 3–1, or at best 5–1. This low turndown ratio may necessitate the installation of two or more units in order to secure the required flexibility and process control. A wide range of sizes is commercially available in pressure CO_2 generators. This commercial equipment is well-designed, reliable, and includes all auxiliaries and safety controls.

Submerged combustion of natural gas is another method for CO_2 generation. Air and natural gas are compressed and then burned under water at the point of application, that is, in the recarbonation basin. Automatic underwater electric ignition equipment is used to start combustion. Submerged combustion is a simple, efficient means of CO_2 generation which provides good control of recarbonation, and requires a minimum of maintenance. The turndown ratio of this type of burner is only about 2–1, so that it is necessary to provide enough burner assemblies to obtain the desired range of control in the amount of CO_2 applied.

LIQUID CARBON DIOXIDE

Commercial liquid CO_2 has found increasing use for recarbonation in water softening plants in the last few years, primarily because of its steadily decreasing cost. However, the price of liquid CO_2 depends greatly on the distance from the source of supply, and the first factor to be investigated is the cost of liquid CO_2 delivered to the plant under consideration. Current (1970) bulk liquid CO_2 prices delivered to points near a source of supply range from \$25 to \$50 per ton. Even in favorable locations, high cost is still the principal disadvantage of using liquid CO_2. Its advantages include flexibility, ease of control, high purity and efficiency, and the smaller piping required because of its high CO_2 content of 99.5 percent as compared to the 6–18 percent as obtained from other sources.

Liquid CO_2 may be delivered to customers in insulated tank trucks ranging from 10–20 tons in capacity. Rail car shipments of 30–40 tons are available to large-volume users. Some manufacturers will lease tank cars so that they may be used for storage at the site, thus eliminating the need and expense of bulk storage tanks and auxiliaries at the plant. For small plants, liquid CO_2 is also available in 20–50 lb cylinders.

Bulk storage tanks may be purchased or leased. Capacities range from 1–100 tons, although the common sizes are 4 through 48 tons. Storage tanks must be insulated and equipped with freon refrigeration and electric or steam vaporization equipment. The working pressure for storage tanks is 350 psi, and the ASME Code for Unfired Pressure Vessels requires hydrostatic testing to 525 psi, or 1.5 times the working pressure. The tanks may be insulated with pressed cork or polyurethane foam. The cooling and vaporizing systems are designed to maintain the liquid CO_2 at about 0°F and 300 psi. If temperature and pressure rise, the cooling system comes on, and on falling pressure, the vaporizer comes into service. In the event either of these systems fails, or, in the event of fire or other accident, the storage tanks are fitted with high- and low-pressure alarms, two safety pop valves, a manual bleeder relief valve, and a bursting disc.

Either liquid or gas feed systems may be used to apply the liquid CO_2 to the wastewater to be treated. In withdrawing CO_2 from the storage tank the pressure is reduced. This pressure reduction cools the CO_2, with the danger of dry ice formation if the expansion is too rapid. Consequently, it is common practice to reduce the pressure in two stages from the 300 psi tank pressure to the 20 psi pressure ordinarily required for feeding the CO_2. Vapor heaters may also be used just ahead of the pressure reducing valves.

For CO_2 gas feed, an orifice plate in the feed line with a simple manometer may be used to measure flow, and a manual valve installed downstream may be used to regulate or control the amount of CO_2 applied. Automatic control to a manual set point can be provided by using a differential pressure transmitter on the feed line orifice, and connecting it to an indicating controller which would operate the control valve. Optionally, the CO_2 feed could be made fully automatic by providing *p*H control. In this case, an electrode would be installed to measure the *p*H of the recarbonated water. This signal would be amplified and sent through a controller which would throttle the control valve on the feed line at low *p*H and open it at high, as set on the controller.

For solution feed of CO_2, equipment similar (except for materials of construction) to solution feed chlorinators may be used, Chlorinator capacity is reduced about 25 percent when feeding CO_2. Approximately 60 gal of water are required to dissolve 1 lb of CO_2 at room temperature and atmospheric pressure. Absorption efficiency with solution feed of CO_2 approaches 100 percent.

CARBON DIOXIDE PIPING AND DIFFUSION SYSTEMS

Because in recarbonation systems the gas temperature is usually in the range of 70–100°F, and pressure is about 6–8 psi, and because CO_2 pipe runs are usually less than 100 ft, it is convenient to use Table 5-1 to estimate the pipe size required.

Table 5-1 Approximate Gas Carrying Capacity.

Pipe Diameter, In.	*Capacity, cfm*
1	45
2	250
3	685
4	1,410
6	3,870

The pipe sizes obtained from Table 5-1 are, of course, an approximation. For greater accuracy, or for long lines, the following modification of the Darcy-Weisbach formula may be used for pressure loss in air piping:

$$\Delta p = \frac{f}{38{,}000} \frac{LTQ^2}{pD^5}$$

where Δp = pressure drop in psi

$$f = \frac{0.048\ D^{0.027}}{Q^{0.148}}$$

(Note: usual values for $f = 0.016–0.049$)

L = pipe length in feet

T = absolute F temperature of the gas = F° + 460

Q = gas flow in cfm

P = absolute pressure of the gas in psi (or line pressure in psi + 14.7)

D = pipe diameter in inches

The pressure loss in elbows and tees can be approximated by use of the following formula:

$$L = \frac{7.6\ D}{1 + \frac{3.6}{D}}$$

where L = equivalent length of straight pipe in feet
D = pipe diameter in inches

The loss in globe valves is about

$$L = \frac{11.4\ D}{1 + \frac{3.6}{D}}$$

where L and D are the same as above.

Carbon dioxide absorption systems often consist of a grid of perforated pipe submerged in the wastewater. The recommended minimum depth of submergence is 8 ft. With lesser depths of submergence some undissolved CO_2 will escape at the water surface. Properly designed absorption systems will put into solution 85–100 percent of the applied CO_2. PVC pipe is an excellent material for the perforated CO_2 grid pipes. Current practice is to use 3/16-in.-diameter orifices drilled in the bottom of the pipe at an angle of 30 deg to the right of the vertical centerline, then 30 deg to the left alternating at a spacing of about 3 in. along the centerline of the pipe. Another arrangement is to point the orifices straight up at the top of the pipe and to direct a jet of water from a header down at the CO_2 orifice, in order to form fine bubbles of the gas which dissolve more readily in the water. Since PVC does not corrode under acid conditions, the openings are not subject to plugging as they are in metal pipes. If 3/16-in. orifices are used, each opening is often rated at 1.1–1.65 cfm, which corresponds to head losses through the orifice of 3 and 8 in. of water column respectively. This is sufficient loss through the orifice to insure good distribution of the CO_2 to each opening. CO_2 laterals must be laid with the same depth of submergence on each orifice. If the size of pipe changes, then eccentric reducers should be used to keep the bottom of the pipe level (assuming that the holes are in the bottom of the pipe). Horizontal spacing between CO_2 diffusion laterals should be at least 1.5 ft in order to get good absorption. To convey cool, dry CO_2, plain steel or cast-iron pipe may be used, but for hot, moist CO_2 gas, the use of stainless steel or other acid-resistant metal is suggested. Special pipe is also required to convey liquid CO_2 in water;

a 1.5-in. cotton fabric hose with openings of controlled size or porosity has been used successfully. Basin hydraulics must take into account raised water levels caused by CO_2 injection.

CARBON DIOXIDE REACTION BASINS OR INTERMEDIATE SETTLING BASINS

Contrary to many early reports in the literature, the recarbonation reaction is not instantaneous. Although about 90 percent of the applied CO_2 does enter the water as dissolved CO_2 in its very short upward journey from the distribution grid through 8 ft of water to the water surface, the time for complete reaction between the dissolved CO_2 and hydroxide and carbonate ions may be as great as 15 minutes in cold water. In the primary phase of two-stage recarbonation, if the reaction is allowed to go to completion at a *p*H near 9.3, the calcium carbonate formed is not redissolved in the second phase of recarbonation to a low *p*H—say to *p*H = 7.0. Magnesium salts do tend to redissolve under these conditions. In the recarbonation of domestic wastewaters, a dense, rapidly settling floc is formed following first-stage recarbonation. This is a rich source of calcium carbonate from which lime (CaO) can be reclaimed and reused by recalcining at temperatures of about 1850°F. In this case, then, it is desirable to allow not only for reaction time (15 minutes) but for enough time to provide some separation of the calcium carbonate by settling. This will require a settling basin with at least 30 min detention at maximum flow rate, and a basin surface overflow rate of not more than 2,400 gal/ft^2/day. This intermediate settling basin should be fitted with continuous mechanical sludge removal equipment. Figure 5-5 shows a two-stage recarbonation system with intermediate reaction and settling for a design flow of 7.5 mgd.

Single-stage recarbonation systems should be followed by 15 min of detention for completion of the chemical reactions, but no provisions for settling or sludge collection are required. The light, cloudy floc which may be formed at times with single-stage recarbonation is removed quite readily by mixed-media filtration with little effect on filter effluent turbidity, head loss, or length of filter run.

Recarbonated lime-treated wastewater should not be applied directly to beds of granular activated carbon witout filtration, because, even at low *p*H (say 7.0) there is still sufficient deposition of calcium carbonate to cause serious problems in the carbon treatment, which easily can be avoided by prior filtration.

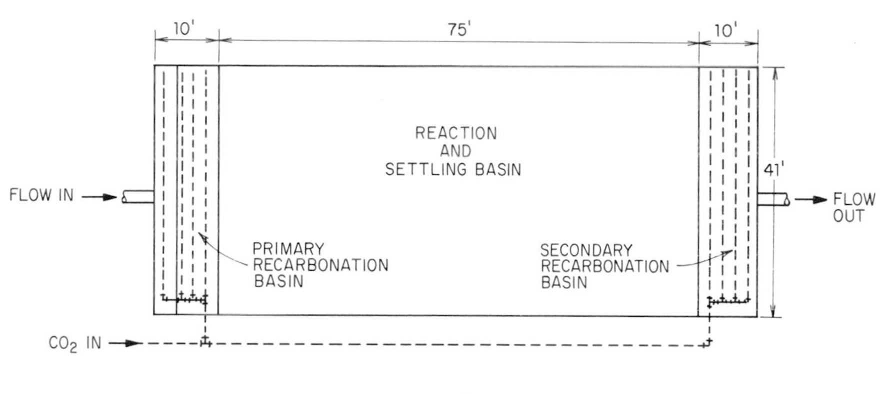

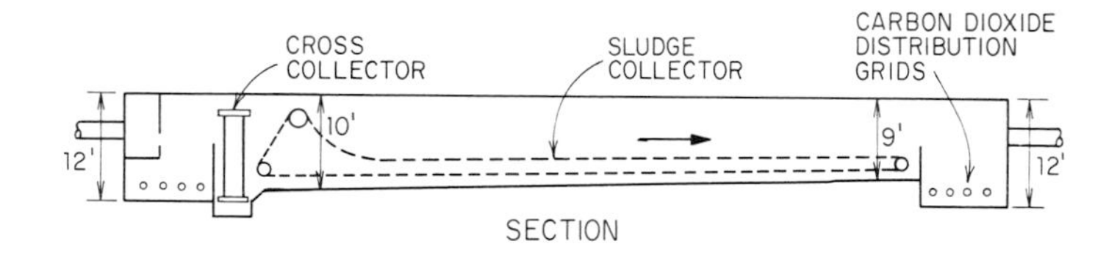

Figure 5–5 Plan and section showing two-stage recarbonation basin with intermediate settling for a 7.5-mgd plant.

OPERATION AND CONTROL OF RECARBONATION

The operation and control of recarbonation systems is easy and simple. Automated control systems ordinarily use a single point of *p*H measurement following the last stage of recarbonation as the basis of control. In two-stage recarbonation systems, the split of total CO_2 flow between the two stages of treatment is fairly constant once it is established for a given flow and the particular set of *p*H values desired, and control based on the final *p*H alone is satisfactory without readjustment of the valves supplying the first and second stage CO_2 supply headers. An indirect but more sensitive control of recarbonation is provided by alkalinity measurements. Continuous reliable automatic monitoring and control equipment is available for either the *p*H or alkalinity method, but the alkalinity measuring equipment is considerably more expensive than the *p*H equipment. Manual control is also quite satisfactory, based either on grab sampling and analysis, or from observation of continuous automatic monitoring of *p*H or alkalinity of the recarbonated water. The CO_2 demands do not vary rapidly or widely, and manual control of dosage is better than might be expected.

SAFETY

Under certain conditions, carbon dioxide can be dangerous and there are safety precautions which must be observed. Prolonged exposure to concentrations of 5 percent or more CO_2 in air may cause unconsciousness and death. The maximum allowable daily exposure for a period of 8 hours is 0.5 percent CO_2 in air. Carbon dioxide is 1.5 times as dense as air, and therefore will tend to accumulate in low, confined areas. Filter-type gas masks are not useful in atmospheres containing excess CO_2, and self-contained breathing apparatus and hose masks must be used. Contact with liquid CO_2 by the skin can cause frostbite. Recarbonation basins must be located out-of-doors and enclosed structures must not be built above them, because of the danger of excessive amounts of CO_2 accumulating within the structures. Before repairmen enter recarbonation basins, the CO_2 supply should be turned off and the space thoroughly ventilated.

In the use of liquid CO_2 there are many other safety considerations too numerous and detailed to be covered completely here. Complete published information can be obtained from liquid CO_2 suppliers or from the Compressed Gas Association, Inc., 500 Fifth Ave., New York, N.Y.

References

1. Anonymous, "Submarine Burners Make CO_2 for Softening Recarbonation," *Water Works Engineering,* 1963, p. 182.
2. Bulletin No. 7–W–83, "Carball CO_2 for Recarbonation," Walker Process Co. (May, 1966).
3. Compressed Gas Assoc., *Handbook of Compressed Gases,* Van Nostrand Reinhold Co., New York, 1962.
4. Fair, M. F., and Geyer, J. C., *Water Supply and Waste Disposal,* John Wiley & Sons, New York, 1954.
5. Haney, P. D. and Hamann, C. L., "Recarbonation and Liquid Carbon Dioxide," *Journal American Water Works Assoc.,* 1969, p. 512.
6. Hoover, C. P., *Water Supply and Treatment,* 8th ed. Bulletin 211, National Lime Association, Washington, D.C.
7. Pamphlet G-6, "Carbon Dioxide," 2nd ed., Compressed Gas Assoc., New York, 1962.
8. Pamphlet G-6.1T, "Tentative Standard for Low Pressure Carbon Dioxide Systems at Consumer Sites," Compressed Gas Assoc., New York, 1966.
9. Ross, R. D., *Industrial Waste Disposal,* Van Nostrand Reinhold Co., New York, 1968.
10. Scott, L. H., "Development Of Submerged Combustion For Recarbonation," *Journal American Water Works Assoc.,* 1940, p. 93.

6
Filtration

GENERAL CONSIDERATIONS

The goal of filtration is the removal of particulate matter. Before selecting the filtration process to be used, it is necessary to accurately determine the total amount of filterable solids, the variability of their quantity, the fraction to be removed, and their characteristics.

The filtration characteristics of the solids found in a biological treatment plant effluent are greatly different than the floc resulting from chemical coagulation for phosphate removal, for example. When no coagulants are used, the filterability of solids in a biological plant effluent is dependent upon the degree of flocculation achieved in the biological process. A trickling filter achieves only a poor degree of flocculation and efficient filtration of the effluent from a trickling filter plant will usually provide only about 50 percent removal or less of the suspended solids normally present. The activated sludge process is capable of much higher degrees of biological flocculation than is the trickling filter process. Culp and Hansen (1967) found that up to 98 percent of the suspended solids found in an extended aeration plant effluent with 24 hr aeration of domestic sewage could be removed by filtration to produce turbidities as low as O.3 JU without the use of coagulants. These authors later reported (June, 1967) that pilot plant studies showed that the degree of biological flocculation achieved in an activated sludge plant was directly proportional to the aeration time and inversely proportional to the ratio of the amount of organic material added per day to the amount of suspended solids in the aeration chamber (load factor). Variation of mixed liquor suspended solids in the normal operating range of 1,500–5,000 mg/l did not significantly affect the filterability of the effluent at a given aeration time and load factor. For domestic wastes, aeration times of 10 hr or more were found to provide flocculation adequate to permit an efficient downstream filter to remove 90–98 percent of the effluent suspended solids. The flocculation provided by aeration times of 6–8 hr. with domestic wastes enabled 70–85 percent suspended solids removal from the secondary effluent.

The strength of biological floc is much greater than that of floc resulting from chemical coagulation. Tchobanoglous and Eliassen (1970) have also noted this basic difference. As a result, biological floc can be removed with coarser filters at higher filtration rates than can the weaker chemical flocs which tend to shear and penetrate a filter more readily. Chemical floc strength can be controlled to some degree with the use of polymers as coagulant aids.

The types of filtration processes discussed in this book can be placed in two general categories: (1) surface filtration devices, including microscreens, diatomaceous earth filters, slow sand filters, rapid sand filters, and moving bed filters; and (2) in-depth filtration devices—granular media filters graded coarse to fine in the direction of flow.

SURFACE FILTRATION

Microscreening

Microscreens are mechanical filters which consist of a horizontally mounted drum, the cylindrical surface of which is made up of a special metallic filter fabric, and which rotates slowly (peripheral speeds of up to 100 fpm) in a tank with two compartments so that water enters the drum from one end and flows out through the filtering fabric. The drum is usually submerged to approximately two-thirds of its depth. The solids are retained on the inside of the rotating screen and are washed from the fabric by pumping strained effluent through a row of jets fitted on top of the machine. Figures 6-1, 6-2, and 6-3 illustrate typical microscreening equipment. The wash water containing the solids flushed from the screen is collected

Figure 6–1 Microscreen installed in concrete tank. *(Courtesy Crane-Cochrane Co.)*

in a hopper or trough inside the drum for return to the secondary plant. The volume of wash water may vary from 3–5 percent and contain 700–1,000 mg/l of suspended solids. The rate of flow through the microscreen is determined by the head applied (normally limited to about 6 in. or less) and by the concentration and nature of the suspended solids in the influent. Typical loading rates are in the range of 2.5–10 gpm/ft^2 of filtering fabric.

The fabrics are usually made up of woven stainless steel of FSL quality (type 316, 18/8) containing 2–3 percent molybdenum (Diaper, 1968). Three aperture sizes are offered by one manufacturer: 60, 35, and 23 microns.

Tertiary Filtration with Microscreens

The operating results when applied to secondary effluents have been reported in several publications with somewhat mixed opinions. The finest fabric (23 microns) was found to remove 89 percent of the suspended solids from an activated sludge plant effluent in studies at Lebanon, Ohio (Bodien and Stenburg, 1966). The 35μ screen removed 73 percent of the suspended solids. The percentage of backwash water averaged 5 with a range of 3–23 percent. The

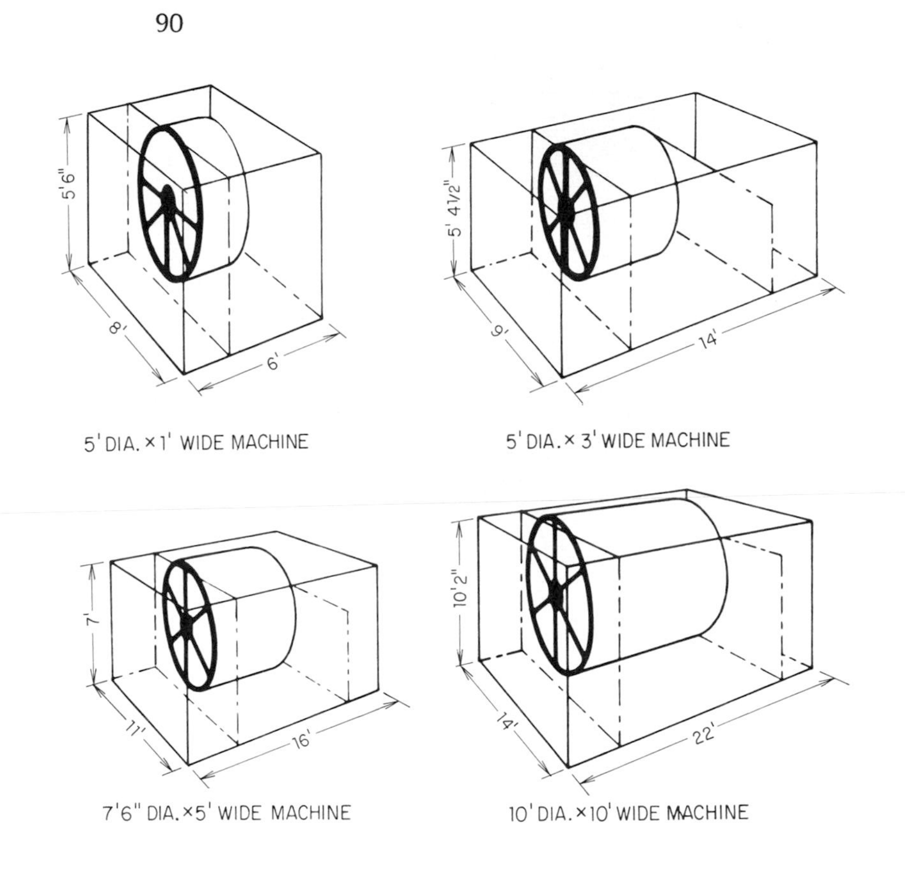

Figure 6–2 Minimum internal dimensions of microscreen tanks. *(Courtesy Crane-Cochrane Co.)*

major operating problems noted at Lebanon were fouling of the screen with grease and the reduction of throughput during periods of high solids carryover from the secondary plant. In discussing the Lebanon tests, Convery (1968) reported that an increase in influent solids from 25–200 mg/l resulted in a decrease in throughput rate from 60–13 gpm. This sensitivity to variations in solids loadings is cited frequently in the literature as a major weakness in the use of microscreening techniques for tertiary solids removal. In an attempt to overcome this problem, a system of automatic control has been devised by the manufacturers. The head loss across the screen is measured pneumatically and transmitted to a converter mechanism which sends a signal to increase the speed of the drum and backwashing pressure on rising head loss. However, no quantitative data defining the allowable variations in solids loading with this modification were available at the time of this writing.

Figure 6–3 5 x 3 ft microscreen in steel package unit. *(Courtesy Crane-Cochrane Co.)*

Diaper (1968) suggests the following guides on anticipated removals from secondary effluents:

Fabric Aperture	*Anticipated Removal (%)* *Suspended solids*	*BOD*	*Flow (gpm/ft² of submerged area)*
23 microns	70–80	60–70	6.7
35 microns	50–60	40–50	10.0

Extensive tests at the Chicago Sanitary District were reported by Lynam, *et al.* (1969). They report about 3 percent backwash water at 3.8 gpm/ft² hydraulic loading. Their data indicate that a final effluent suspended solids of 6–8 mg/l and a BOD of 3.5–5 mg/l could be expected when applying a good quality (20–25 mg/l suspended solids and 15–20 mg/l BOD) activated sludge effluent to a 23μ microscreen. They also noted that the microscreen was more responsive to suspended solids loading than to hydraulic loading and that the maximum capacity of the microscreen was reached at the loading of 0.88 lb/ft²/day at 6.6 gpm/ft². Unfortunately, no solids capacity data were developed for lower hydraulic loading rates.

Diaper (1968) reports that microscreens (35μ aperture) have reduced the secondary effluent solids from 17 mg/l–5.7 mg/l, on the average, for a trickling filter located at Berkshire, England. Tests at Brampton, Ontario, showed BOD and suspended solids reductions of 53.5 and 57 percent respectively when microscreening an activated sludge effluent. Evans (1957) reports that microscreening of the secondary effluent from the Luton, England, plant (activated sludge followed by trickling filter treatment) produced a final effluent with an average BOD of 6.6 mg/l and suspended solids of 6.3 mg/l. Dixon and Evans (1966) reported application of a microscreen to a trickling filter plant effluent which contained 36–220 mg/l suspended solids. The 23μ microscreen effluent contained a 15–40 mg/l. The authors felt that a 20 mg/l BOD and 20 mg/l suspended solids requirement could be generally met by such a system. The somewhat varied results noted above are no doubt related to the variations in degree of biological flocculation achieved by the various secondary plants involved. The other references cited above indicate that microscreening of well-flocculated biological solids, as found in some activated sludge plant effluents, can provide final effluent solids of less than 10 mg/l, in cases where the applied suspended solids are consistently less than 35 mg/l.

Clarification of Unsettled Trickling Filter Effluent

Truesdale and Birkbeck (1966) have applied a 35-micron microscreen directly to the effluent from the second-stage trickling filters at the two-stage trickling plant at the Harpenden (England) sewage works. They reported that the performance of the microscreen was consistently superior to that of a parallel settling basin operating at an overflow rate of about 500 gpd/ft^2. The daily average concentrations of suspended solids in the effluents from the trickling filters, settling basin, and microscreen were 46.9, 22.3, and 12.8 mg/l respectively. The maximum throughput was found to be 2.5–6 gpm/ft^2 of filter fabric. The suspended solids data showed that the concentration of suspended solids was greater than 20mg/l only 6 percent of the time. They then attempted to apply the microscreen to the primary trickling filter effluent but found that the microscreen throughput fell by nearly 70 percent in a few days due to heavy slime growths. Ultraviolet lamps and the use of hypochlorite for cleaning gave only temporary improvement. They concluded that the primary trickling filter effluent would not be amenable to microscreening. It has been reported by Diaper (1969) that the promising results obtained on direct microscreening of secondary trickling filter effluent has led to the design of a permanent installation at Fleet, Hampshire, England, where the microscreens may take filter effluent directly. The results from this installation should enable the evaluation of the feasibility of supplementing or replacing two-stage trickling filter plant secondary clarifiers with microscreens.

Clarification of Chemically Coagulated Sewage

Lynam, *et al.* (1969) report that microscreens cannot effect filtration of alum coagulated secondary effluent solids. This is not surprising since the low strength of such chemical floc noted earlier leads to rapid shearing and penetration of the microscreen. This fact does limit the flexibility of a microscreening installation. Even if the microscreen presents a suitable solution to an immediate suspended solids and BOD removal problem, the microscreen could not be used for removal of chemical floc should such chemical treatment be required for phosphorus removal in the future.

Design of Microscreening Installations

There is no simple means such as dividing total flow rate by suitable hydraulic filter loading rate to determine the necessary filter area. The hydraulic capacity of a microscreen is governed by the rate of clogging of the fabric, the rotational speed of the drum, the area of submerged screen, and the available head. Boucher (1947) derived a formula for the hydraulics of automatic strainers based on the concept of the filterability of fluids:

$$H = \frac{m\ Q\ C_f e^{hIQ/s}}{A}$$

where H = head loss across microscreen (inches)

Q = constant, total rate of flow through the unit (gpm)

C = initial resistance of the clean filter fabric (feet) at a given temperature and standard flow conditions. (23μ microscreen $C_f = 1.8$ ft; 35μ microscreen $C_f = 1.0$ ft)

I = filterability index of influent measured on fabric in use (an expression of the volume of water obtained per unit head loss when passed at a standard rate through a unit area of standard filter)

S = speed of strainer expressed as number of square feet of effective fabric entering water in given time (square feet per minute)

A = Effective submerged area (square feet)

Constants: $m = 0.0267$; $n = 0.1337$

Figure 6–4 Portable filtrameter used for predicting size and efficiency of microscreens. *(Courtesy Crane-Cochrane Co.)*

The filterability index can be determined in a number of ways. The most accurate method actually measures the head loss by means of a mercury manometer across a section of filter being tested using the feed in question with the flow controlled by a constant head box. A portable instrument (Figure 6-4) is also available.

Bodien and Stenburg (1966) determined the filterability index for the activated sludge effluent used in their studies at Lebanon. Their data showed the following relationships: $I = 0.494c$ where c is equal to the concentration of influent suspended solids in mg/l. They determined an average value of $I = 17$ for the 23μ screen and calculated the capacity of a 5×1 ft screen with a maximum headloss of 6 in. and a maximum peripheral drum velocity of 508 fpm for S as 58 gpm or .0835 mgd. The following table summarizes the microscreen sizes available from Glenfield and Kennedy:

Drum Sizes (feet) diam.	width	*Motors (bhp)* drive	wash pump	*Approx. Ranges of Capacity (mgd)*	*Recommended Maximum Flow for Tertiary Sewage Applications (mgd)* 23μ	35μ
5	1	½	1	0.05 to 0.5	0.075	0.11
5	3	¾	3	0.3 to 1.5	0.20	0.30
7½	5	2	5	0.8 to 4	0.70	1.00
10	10	4	7½	3 to 10	2.00	3.00

Another manufacturer (Zurn Industries) offers the following units:

Drum Sizes (feet) *diam.*	*width*	*Screen Area (sq ft)*
4	2	24
4	4	48
6	4	72
6	6	108
6	8	144
10	10	315

Diaper (1968) offers the following estimates of installed costs; including all mechanical equipment, installation, building, and civil engineering work:

23μ aperture screen, 7.5 × 5ft units up to 3 mgd flow, $57,500/mgd
35μ aperture screen, 7.5 × 5ft units up to 3 mgd flow, $40,000/mgd
23μ aperture screen, 10 × 10ft units up to 50 mgd flow, $40,000/mgd
35μ aperture screen, 10 × 10ft units up to 50 mgd flow, $28,000/mgd

He estimates the operating costs for the 23μ screens at $6/mg and for the 35$\mu$ screen at $4.50/mg including power, supervisory labor, and maintenance.

Smith (1968) has also estimated the costs (adjusted to June, 1967) of microscreening installations, which are summarized below:

Plant Capacity (mgd)	*Capital Costs (dollars)*	*Operating & Maint. (dollars/mg)*
1	54,000	8.00
3	150,000	7.25
5	230,000	7.75
10	420,000	6.50
25	1,000,000	5.90
50	1,800,000	5.40

Diatomaceous Earth Filtration

Several investigators have studied diatomaceous earth (D-E) filtration of secondary effluents. D-E is fed at a controlled rate to the secondary effluent which is then passed through a precoated filter septum. Shatto (1960) found D-E filtration to produce a final effluent with "no detectable BOD and only a trace of suspended solids." However, the ability of a D-E filter to produce an excellent quality effluent is accompanied by extremely high cost. A major problem

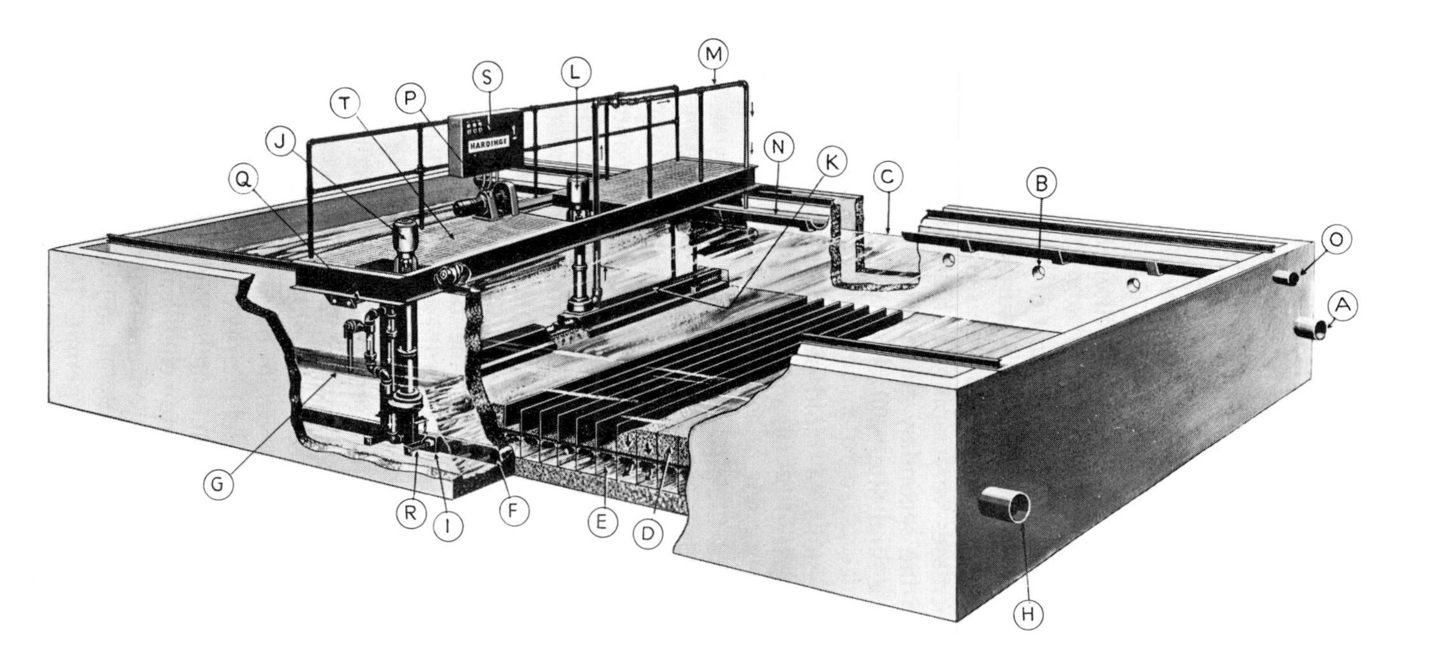

Figure 6–5 The Hardinge automatic backwash sand filter. *(Courtesy Koppers Co.)*

A. Influent line.
B. Influent ports.
C. Influent channel.
D. Compartmented sand filter bed.
E. Sectionalized under-drain.
F. Effluent and backwash ports.
G. Effluent channel.
H. Effluent discharge line.
I. Backwash valve.
J. Backwash pump assembly.
K. Washwater hood.
L. Washwater pump assembly.
M. Washwater discharge pipe.
N. Washwater trough.
O. Washwater discharge.
P. Mechanism drive motor.
Q. Backwash support retaining springs.
R. Pressure control springs.
S. Control instrumentation.
T. Traveling backwash mechanism.

a coarse, shallow filter. The significance of floc strength is well illustrated by the fact that tests conducted with alum coagulation and settling of the effluent prior to filtration showed a slight deterioration compared to filtration without pretreatment by alum coagulation. Lynam and his coauthors felt the poorer performance was due to insufficient coagulation resulting in floc too fine to be removed. Filter performance was improved by the use of an anionic polymer in conjunction with the alum treatment. The authors of this book feel that the improved performance was primarily due to an increase in floc strength rather than to an increase in floc size. A filter as coarse (0.58 mm effective size) and shallow as that used at Chicago will be very prone to shearing and breakthrough of fragile chemical floc. The data certainly illustrates the point that a coarse filter medium design is better suited for removal of strong biological floc than for removal of fragile chemical floc. The Chicago investigators concluded that the rapid sand filter produced a given degree of treatment more economically than the microscreening device operated in parallel in their study.

Tchobanoglous (1967, 1968) evaluated rapid sand filtration of activated sludge effluent without prior chemical coagulation in a rapid sand filter. As would be expected with a single medium filter, 75–90 percent of the head loss occurred in the upper inch of the rapid sand beds, confirming that such filters are indeed a surface filtration device. At a hydraulic loading rate of 5 gpm/ft^2, a decrease in sand grain size from 1.1–0.5 mm increased the suspended solids removal from 10–40 percent. The activated sludge process was a high-rate process which provided only a limited degree of biological flocculation. The effect of sand size was much more pronounced than was the effect of the rate of filtration for the biological floc being filtered. The lower limit on sand size is determined, of course, by the initial head loss and the rate of head loss buildup with the single medium filter.

The limitations of the single medium rapid sand filter result from its behavior as a surface filtration device. The advantages of in-depth filtration achieved by dual or three media filters can be achieved at lower capital and operating costs than with conventional rapid sand filters. Therefore, if a granular filter is contemplated for use, the filter design selected should provide in-depth filtration rather than merely the surface filtration provided by a rapid sand filter design. Thus, detailed design considerations for granular filters are presented in a following section on in-depth filtration. The conventional rapid sand filter medium offers no advantage over the in-depth filter media designs now available while suffering several severe disadvantages.

The former is not recommended for use in advanced waste treatment plants.

Moving Bed Filter

The moving bed filter (MBF) is a new method of applying sand filtration that is being developed by Johns-Manville Products Corporation. The MBF is a continuous sand filter in which the water moves countercurrent to the sand. The filter medium, sand, is driven through a cone in one direction while simultaneously passing the wastewater to be treated through the filter bed in the opposite direction. The solids are removed from the filter bed as rapidly as is required by their buildup. Movement of the filter bed is accomplished by means of a hydraulically actuated diaphragm. The diaphragm pushes the sand bed as a plug through the cone toward the inlet end. As the diaphragm relaxes, clean sand feeds by gravity into the void left in the bed in front of the diaphragm. The pushing cycle is then repeated. The sludge-sand mixture falls from the face down into the hopper bottom of the head tank, from which it is removed and passed through a washing tower. The cleaned sand is returned to the feed hopper of the MBF. Final effluent from the MBF is used for washing the sand, and to return the sand to the sand hopper. The removal and washing of the sand may be continuous or intermittent. Since the sand is constantly being removed, cleaned, and retured to the system, the filter unit does not have to be stopped for backwashing as do conventional units.

The filter medium usually used is 0.6–0.8 mm sand with a maximum sand feed rate of 12 in./hour. The MBF principle allows much higher solids loadings than permissible with a fixed sand bed. Preliminary results obtained while treating unsettled trickling filter effluent have been reported by Convery (1968). Following alum coagulation directly prior to filtration, influent turbidities were reduced from 22–41 JU to 6.3–10.0 JU, BOD from 40–64 mg/l to 8.8–10.0 mg/l, COD from 111–172 mg/l to 39–43 mg/l, and total phosphate from 30–40 mg/l to 1.5–2.5 mg/l. Polymer dosages of 0.5–1.0 mg/l were applied for the lower filter effluent turbidity values achieved. Plant-scale experiences with the mixed media filters described in a subsequent section have shown that lower turbidities can be achieved in a similar direct coagulation application with substantially lower doses of polymer. It is felt that the coarseness of the media employed in the MBF must be compensated for with high polymer doses to strengthen the floc and prevent its penetration through the filter. The tests to date have clearly shown the MBF can tolerate much higher solids loadings than can a conventional rapid sand filter, a point of considerable merit in sewage filtration applications.

The manufacturer has released technical data (Johns-Manville Engineering Data MBF-2) on the MBF which indicates that the maximum unit capacity is 250,000 gpd/unit. The dimensions of the main unit are about 6 ft wide × 19 ft long × 21 ft 6 in. high, with the auxiliary units occupying a space of approximately the same area. The size and complexity of such a unit would appear to limit the application to plants of relatively small capacity. For example, the operation and maintenance of the 40 units required for a 10 mgd flow would appear to be impractical. The wastewater flow resulting from the sand cleaning operation is listed as 15 gpm for the 200 gpm influent flow of the 250,000 gpd unit, or 7½ percent of the influent flow. The total motor horsepower is listed at 10 for the unit with a total unit weight of 78,000 lb. Capital and operating costs were not available at the time of this writing.

IN-DEPTH FILTRATION

General Theory

The previously discussed sensitivity of a rapid sand filter to high suspended solids concentrations can be readily understood by examination of Figure 6-6A, which is a cross section of a typical single medium filter, such as a rapid sand filter. During filter backwashing, the sand grades hydraulically, with the finest particles rising to the top of the bed. As a result, most of the material removed by the filter is removed at or very near the surface of the bed. Only a small part of the total voids in the bed are used to store particulates and head loss increases very rapidly. When the secondary effluent contains relatively high solids concentrations, a sand filter will blind at the surface in only a few minutes. Typically, 75–90 percent of the head loss occurs in the upper inch of rapid sand beds when filtering activated sludge plant effluent.

One approach to increasing the effective filter depth is the use of a dual media bed using a discrete layer of coarse coal above a layer of fine sand, as shown in Figure 6-6B. The work area is extended, although it still does not include the full depth of the bed, for there is some fine to coarse stratification within each of the layers, as shown by the graph depicting pore size. Effective size of the sand

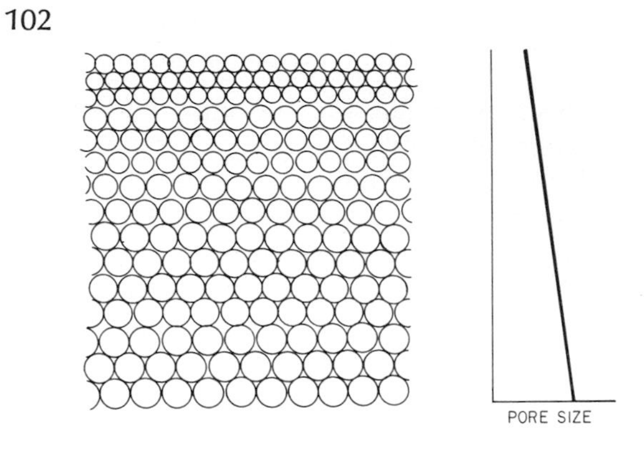

(A)
CROSS-SECTION THROUGH SINGLE-MEDIA BED
SUCH AS CONVENTIONAL RAPID SAND FILTER.

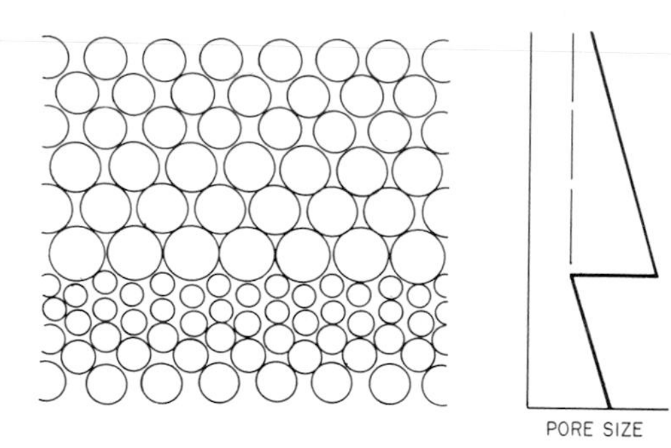

(B)
CROSS-SECTION THROUGH DUAL-MEDIA BED
COARSE COAL ABOVE FINE SAND

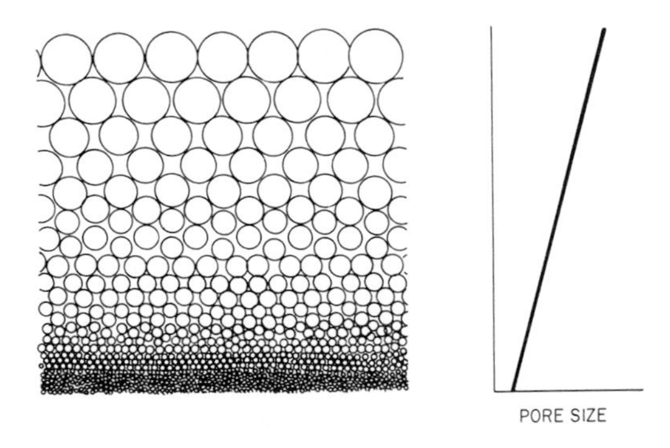

(C)
CROSS-SECTION THROUGH IDEAL FILTER
UNIFORMLY GRADED FROM COARSE TO FINE
FROM TOP TO BOTTOM

Figure 6–6 Graphical representation of various media designs. *(Courtesy Neptune Microfloc, Inc.)*

in a typical dual media filter is 0.4–1.0 mm with a coal size of 0.8–2.0 mm with the layers 10 and 20 in. deep, respectively.

In designing a dual media bed, it is desirable to have the coal (specific gravity about 1.6) as coarse as is consistent with solids removal to prevent surface blinding but have the sand (specific gravity about 2.6) as fine as possible to provide maximum solids removal. However, if the sand is too fine in relation to the coal, the former will actually rise above the top of the coal during the first backwash and remain there when the filter is returned to service. For example, if a 0.2-mm sand were placed below 1.0-mm coal, the materials would actually reverse during backwash with the sand becoming the upper layer and the coal, the bottom. Although the sand has a higher specific gravity, its small diameter in this case would result in its rising above the coal at a given backwash rate. The only way to enable very fine silica sand to be used in the bottom filter layer would be the use of finer coal, which would defeat the purpose of the upper filter layer since the fine coal would be susceptible to surface blinding. Experience has shown that it is not feasible to use silica sand smaller than about 0.4 mm because smaller sand would require coal small enough to result in unacceptably high head loss at rates above 3 gpm/ft^2.

To overcome the above limitation and to achieve a filter which very closely approaches an ideal one (Figure 6-6C), the mixed-media concept was developed (patents held by Neptune Microfloc, Inc., Corvallis, Oregon). The problem of keeping a very fine medium at the bottom of the filter is overcome by using a third, very heavy (garnet, specific gravity of about 4.2, or ilmenite, specific gravity of about 4.5) very fine material beneath the coal and sand. The garnet (or ilmenite), sand, and coal particles are sized so that intermixing of these materials occurs and no discrete interface exists between the three. This eliminates the stratification illustrated for the dual media filter in Figure 6-6B and results in a filter which very closely approximates the idea of a uniform decrease in pore space with increasing filter depth as shown in Figure 6-6C.

Actually, the term "coarse-to-fine" filter refers more accurately to the pore space rather than to the media particles themselves. The illustrations in Figure 6-6 actually represent pore size rather than the size of individual grains. By selecting the proper size distribution of each of the three media, it is possible to construct a bed which has an increasing number of particles at each successively deeper level in the filter. A typical mixed-media filter has a particle size gradation which decreases from about 2 mm at the top to about 0.15 mm at the

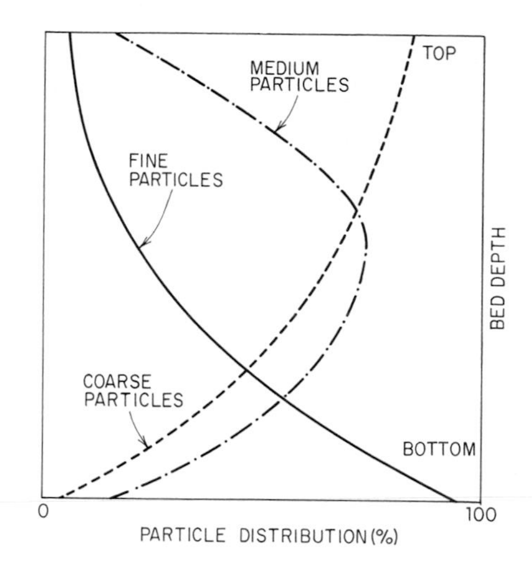

Figure 6–7 Distribution of media in a properly designed mixed-media filter. *(Courtesy Neptune Microfloc, Inc.)*

bottom. The uniform decrease in pore space with filter depth allows the entire filter depth to be utilized for floc removal and storage. Figure 6-7 shows how particles of the different media are actually mixed throughout the bed. At all points in the bed there is some of each component, but the percentage of each changes with bed depth. There is steadily increasing efficiency of filtration in the direction of flow.

There is no one mixed-media design which will be optimum for all wastewater filtration problems. For example, removal of small quantities of high-strength biological floc often found in activated sludge effluents may be satisfactorily achieved by a good dual media design. With a weaker floc strength or with an increase in applied solids loading, the benefits of the mixed, tri-media bed become more pronounced. Comparative data published by Conley (1965) showed that a mixed-media bed will consistently produce longer runs and lower effluent turbidities when removing alum floc at filter rates of 5 gpm/ft^2 than would a dual media bed. Conley and Hsiung (1969) have presented techniques designed to optimize the media selection for any given filtration application. Their work clearly indicates the marked effects that the quantity and quality of floc to be removed can have on media selection (Table 6-1). Removal of the poorly flocculated solids normally found in a trickling filter effluent can be improved by using smaller media than would be used for removal of activated sludge effluent suspended solids. Pilot tests of various media designs can be more than justified by improved plant performance in most cases.

Table 6-1 Illustrations of Varying Media Design for Various Types of Floc Removal *(Conley and Hsiung, 1969).*

Type of Application	*Garnet*		*Silica Sand*		*Coal*	
	size	*depth*	*size*	*depth*	*size*	*depth*
Very heavy loading of fragile floc	− 40 + 80	8 in.	− 20 + 40	12 in.	− 10 + 20	22 in.
Moderate loading of very strong floc	− 20 + 40	3 in.	− 10 + 20	12 in.	− 10 + 16	15 in.
Moderate loading of fragile floc	− 40 + 80	3 in.	− 20 + 40	9 in.	− 10 + 20	8 in.

Size: − 40 + 80 = passing No. 40 and retained on No. 80 U.S. sieves.

Certainly, use of three media does not in itself insure superior performance as illustrated by the experiences of Oakley and Cripps (1969). Using the coal, sand, and garnet materials readily available to them resulted in a bed which did not have significant advantages over other filter types when filtering secondary effluent. Oakley (1969) reports the tri-media bed was made up of 8 in. of 0.7–0.8 mm garnet, 8 in. of 1.2–1.4 mm sand, and 8 in. of 1.4–2.4 mm coal. The authors' experience indicates that this bed was too shallow and too coarse and that a better media selection for the particular application would have been 3 in. of 0.4–0.8 mm garnet, 9 in. of 0.6–0.8 mm sand, and 24 in. of 1–2 mm coal.

One of the key factors in constructing a satisfactory mixed-media bed is the careful control of the size distribution of each component medium. Rarely is the size distribution of commercially available materials adequate for construction of a good mixed-media filter. The common problem is failure to remove excessive amounts of fine materials. These fines can be removed by placing a medium in the filter, baskwashing it, draining the filter, and skimming the upper surface. The procedure is repeated until field sieve analyses indicate an adequate particle size distribution has been obtained. A second medium is added and the procedure repeated. The third medium is then added and the entire procedure repeated. Sometimes, 20–30 percent of the materials may have to be skimmed and discarded to achieve the proper particle size distribution.

The remainder of this chapter will be devoted to filter system design considerations which are largely independent of the exact media design selected, which must be optimized for each specific application.

PERFORMANCE OF IN-DEPTH FILTERS

Plain Filtration

As previously discussed, the degree of solids removal when filtering secondary effluents without the use of chemical coagulation is dependent upon the degree of biological flocculation achieved in the secondary plant. Plain filtration of primary effluent will provide very little suspended solids reduction due to the colloidal nature of most

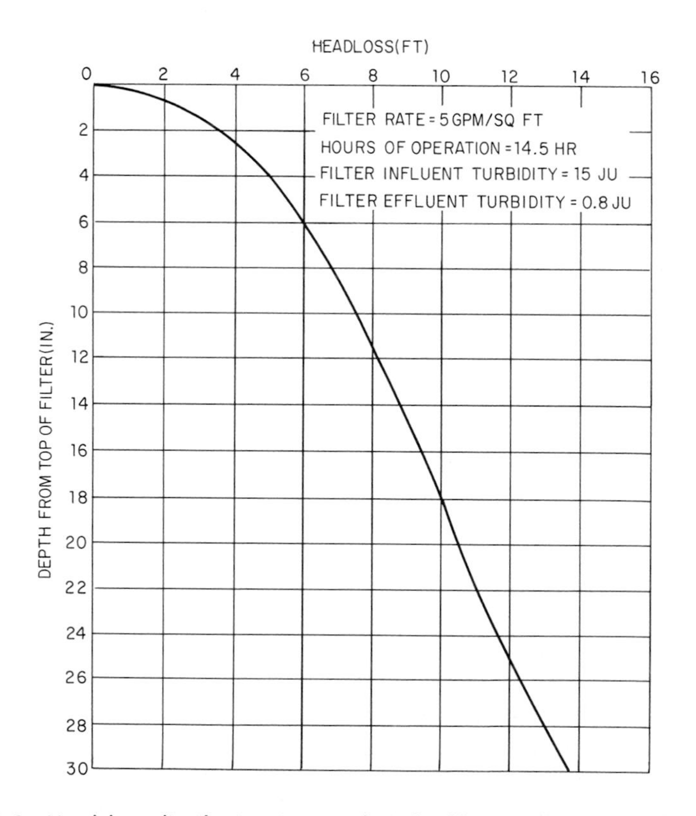

Figure 6–8 Head loss distribution in mixed-media filter applied to activated sludge effluent. (*Courtesy* Water & Sewage Works)

of the solids. Due to the high strength of biological floc, there will be little difference between the effluent quality produced by a well-designed sand filter and a mixed-media filter. However, there will be a very significant difference in favor of the mixed media in length of filter runs, quantity of backwash water, and system reliability.

The advantage of in-depth filtration is well illustrated by the data of Tchobangolous (1969) discussed earlier which show that 90 percent of the head loss occurred in the top 2 in. of a rapid sand filter applied to activated sludge effluent (0.488 mm sand, 5 gpm/ft^2). As shown in Figure 6-8, about 90 percent of the head loss in a mixed-media bed operating under similar conditions occurred in the upper 24 in. of the 30-in. bed, indicating excellent utilization of the bed depth for floc removal (Culp, 1968).

Although a mixed-media filter can tolerate higher suspended solids loadings than can the other filtration processes discussed, it still has an upper limit of applied suspended solids at which economically long runs can be maintained. With activated sludge effluent, suspended solids loadings of up to 120 mg/l, filter runs of 15–24 hr at 5 gpm/ft^2 have been maintained when operating to a terminal head loss of 15 ft of water. Suspended solids concentrations of 300 mg/l or more will lead to uneconomically short filter runs, even when using a mixed-media filter. Figure 6-9 illustrates the effects of influent solids on rate of headloss buildup. The media design shown in Figure 6-9 is somewhat finer in size than that now recommended for plain filtration of activated sludge solids by the media supplier but the general magnitude of head loss buildup can be determined from this curve. Should the secondary plant involved have a history of frequent, severe upsets resulting in secondary effluent suspended solids concentrations of 200–500 mg/l, an intermediate settling tank between the secondary clarifier and the filter with provision for chemical coagulation during upset periods should be made.

Due to the variations in biological flocculation achieved in secondary processes, it is impossible to present accurate estimates of the efficiency of plain filtration without knowing the specific biological process loading parameters. The following quality of filter effluents are presented as general guides to the suspended solids concentrations which might be achieved when filtering a secondary effluent of reasonable quality without chemical coagulation: high-rate trickling filter, 10–20 mg/l; two-stage trickling filter, 6–15 mg/l; contact stabilization, 6–15 mg/l; conventional activated sludge plant, 3–10 mg/l; activated sludge plant with load factor less than 0.15, 1–5 mg/l. The degree of biological flocculation is the limiting factor in determining

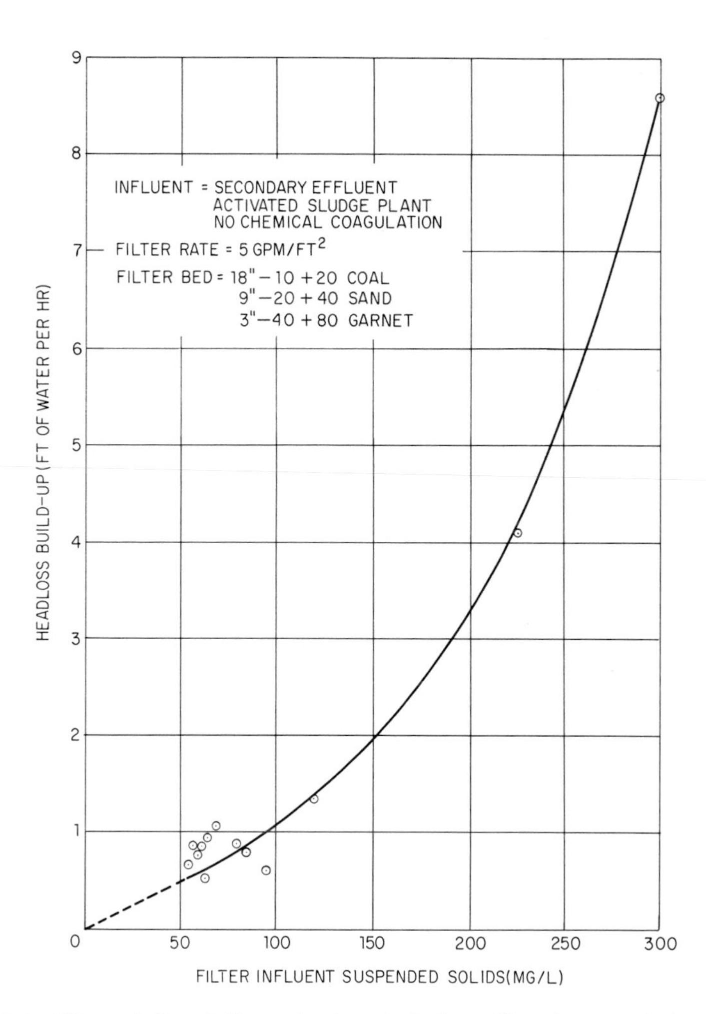

Figure 6–9 Effect of filter influent (activated sludge effluent) suspended solids on head loss buildup for mixed-media filter.

the solids removal. If effluent quality substantially better than that indicated by the general guides above is required, chemical coagulation must be used.

The effluent quality produced by plain filtration of secondary effluents is essentially independent of filter rate within the range of 5–15 gpm/ft² primarily due to the high strength of the biological floc. This statement is based on the results of many tests conducted by the authors which have been confirmed by Tchoganoglous (1968) who reports for his experiments on plain filtration that ". . . for a given sand size, varying the filter rate had little effect on the suspended solids removal characteristics of the filter bed." It must be emphasized that this applies only to removal of strong biological floc and not to the weaker floc resulting from chemical coagulation. Tchobanoglous studied rates of 2–10 gpm/ft^2. The selection of design rate will depend primarily on the anticipated quality of filter influent. Should the operation of the secondary plant be extremely stable and rarely produce secondary effluent suspended solids in excess of 30 mg/l, a filter rate of 10–15 gpm/ft^2 would be satisfactory. Should the plant suffer from inconsistent operation resulting in frequent upsets and occurrences of 50–100 mg/l secondary effluent suspended solids concentrations, a loading rate of 5 gpm/ft^2 would be a better selection. Should the secondary effluent solids frequently exceed 100 mg/l even for short time periods, consideration should be given to providing an intermediate settling basin between the secondary clarifier and the filter, with provisions for chemical coagulation during severe upsets, or to revising the secondary plant design or operation.

Filtration of Chemically Coagulated Effluents

The selection of design rate will be dependent upon the quality of filter influent and the required effluent quality. As noted earlier, the strength of the chemical floc is considerably less than for biological floc with the result that the chemical floc shears more readily, causing filter breakthrough should the filter rate be excessive or the filter media too coarse. Mixed-media filtration of properly coagulated secondary effluent will produce filter effluent essentially free of suspended solids as evidenced by turbidities of 0.1 JU. The filter will also aid substantially in phosphorus removal by removing nonsettleable colloidal material which would contribute to the effluent phosphorus concentration. For example, proper chemical coagulation and settling of secondary effluent can produce a phosphate concentration of 0.5–2 mg/l when operating efficiently. Mixed-media filtration of this effluent is capable of reducing the phosphorus concentration to less than 0.1 mg/l. O'Farrell, *et al.* (1969) report that the dual media filtration step provided 3–14 percent of the overall 90 percent or more phosphorus removal provided by the FWPCA pilot plant at Washington, D.C. At South Tahoe, the filtration step provides 8–12 percent of the overall 99 percent phosphorus removal.

The use of polymers as filtration aids to increase chemical floc strength is discussed in a subsequent section. Such use of polymers will permit higher filter rates and/or coarser filters while reducing the risk of floc shearing.

Filtration rates of 5 gpm/ft^2 are routinely used for application of mixed-media filters for chemical floc removal in potable water treatment. Rates higher than this, up to 10 gpm/ft^2, are practical in most tertiary applications where coagulation and efficient sedimentation precede the filtration step. If the pilot filter study required to optimize the filter rate and media design is not practical, selection of a filter rate of 5 gpm/ft^2 and sizing the filter hydraulics for twice this rate will enable the nominal capacity at 5 gpm/ft^2 to be exceeded should plant-scale tests indicate higher filter rates are permissible.

The fact that mixed-media beds can tolerate high solids loadings permits the application of coagulants directly to the filter influent without presettling. The coagulant doses required for phosphate removal are normally high enough that presettling is economically justified and should be used. However, if clarification only and not phosphate removal is the goal of coagulation, coagulant doses up to 40–50 mg/l with secondary effluent suspended solids loadings of 25–35 mg/l can be handled with direct filtration at rates of 5 gpm/ft^2. If direct filtration of coagulated effluent is considered, a pilot study is mandatory. Conley and Hsiung (1969) have reported on experiences with some of the numerous water treatment plants using direct, mixed-media filtration of coagulated water. They report one case with 12-hr filter runs at 5 gpm/ft^2 with 32 mg/l of alum, 0.2 mg/l of polymer, and 25 units of turbidity applied directly to the filter without prior flocculation or settling. Filtered water turbidities were 0.2–0.4 JU.

Use of Polymers as Filtration Aids

Polymers are high molecular weight, water soluble compounds which can be used as primary coagulants, settling aids, or filtration aids. They may be cationic, anionic, or nonionic in charge. Generally, the doses required for coagulation or as a settling aid in conjunction with another coagulant far exceed that needed as a filtration aid. Typical doses when used as a settling aid are 0.1–2.0 mg/l while doses of less than 0.1 mg/l are often adequate to serve as a filtration aid. When used as a filtration aid, the polymer is added to increase the strength of the chemical floc and to control the depth of penetration of floc into the filter. The polymer is not added to improve coagulation but rather to strengthen the floc made up of previously coagulated material to minimize the prospects of shearing fragile floc, causing filter breakthrough. For maximum effectiveness as a filtration aid, the polymer should be added directly to the filter influent and not in an upstream settling basin or flocculator. However, if polymers are used upstream as settling aids, it may not be necessary to add any additional polymer as filtration aid.

Figure 6-10 illustrates the effects of polymers as filter aids. The conditions represented by Figure 6-10 A illustrate the results of a fragile floc shearing and then penetrating the filter causing a premature termination of its run due to breakthrough of excessively

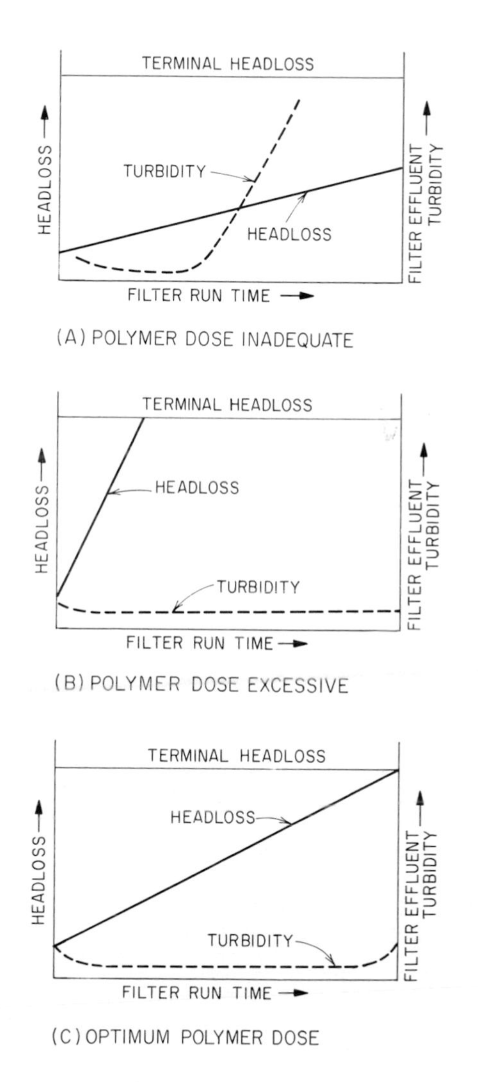

Figure 6–10 Effects of polymers as filtration aids.

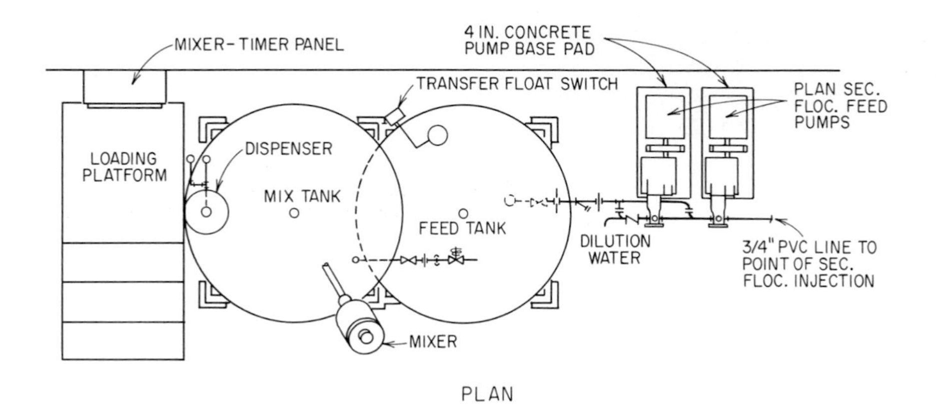

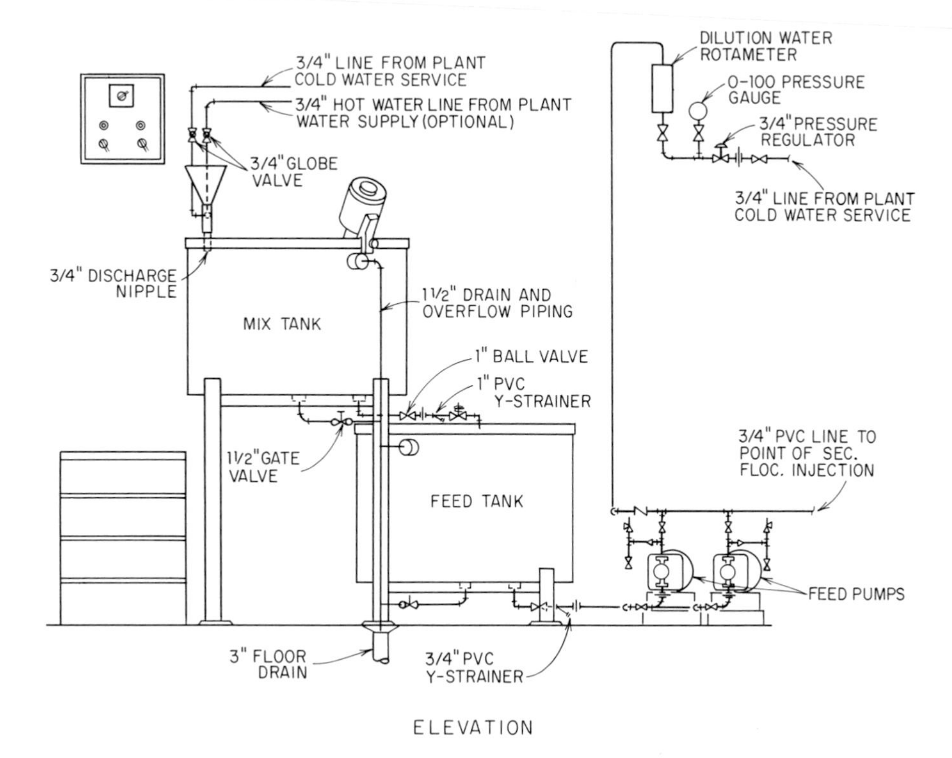

Figure 6–11 Polymer mixing and feed equipment. *(Courtesy Neptune Microfloc, Inc.)*

Figure 6–12 Typical polymer feed station. *(Courtesy Neptune Microfloc, Inc.)*

high effluent turbidity. If the polymer dose is too high (Figure 6-10B), the floc is too strong to permit penetration into the filter causing a rapid buildup of head loss in the upper portion of the filter and a premature termination due to excessive head loss. The optimum polymer dose will permit the terminal head loss to be reached simultaneously with the first sign of increasing filter effluent turbidity (Figure 6-10C).

Many polymers are delivered in a dry form. They are not easily dissolved and special polymer mixing and feeding equipment is required. Typical equipment is shown in Figure 6-11 and 6-12. Many polymers are bio-degradable and cannot be stored in dilute solution for more than a few days without suffering significant degradation and loss of strength.

Types of Filter Structures

The gravity and pressure filter structures commonly used in water treatment plants are readily adaptable to advanced wastewater treatment plant filtration. Pressure filters are often advantageous in waste treatment applications for the following reasons:

1. In many wastewater applications, the applied solids loading is higher and more variable than in a water treatment application. Thus it is desirable to have higher heads (up to 20 ft) available than practical with gravity filter designs to provide maximum operating flexibility.
2. In advanced waste treatment processes, the filtration step is frequently followed by another unit process (carbon adsorption, ion exchange, etc.). The effluent from a pressure filter can be passed through a downstream process without having to pump the filter effluent, often eliminating a pumping step which would be required with a gravity filter.
3. All filter wash water must be treated in sewage applications. The ability to operate to higher head losses with a pressure filter reduces the amount of wash water to be recycled.
4. Pressure filter systems are usually less costly in small and medium-sized plants.

A typical gravity filter section is shown in Figure 6-13. Gravity filter areas usually do not exceed 800–1000 sq ft per filter unit. Two filter units may be combined in one structural cell.

Pressure filters may be either horizontal or vertical. The horizontal filters offer much larger filter areas per unit and would normally be used when plant capacities exceed 1–1.5 mgd.

A typical horizontal pressure filter vessel is shown in Figure 6-14. Although an 8-ft diameter vessel is shown, 10-ft diameters are also commonly used with lengths up to 60 ft. The 10-ft diameter by 38-ft long horizontal pressure filters in use at the South Lake Tahoe Water Reclamation Plant are shown in Figure 6-15.

A typical vertical filter is shown in Figure 6-16. Diameters up to 11 ft are commonly used with working pressures up to 150 psig.

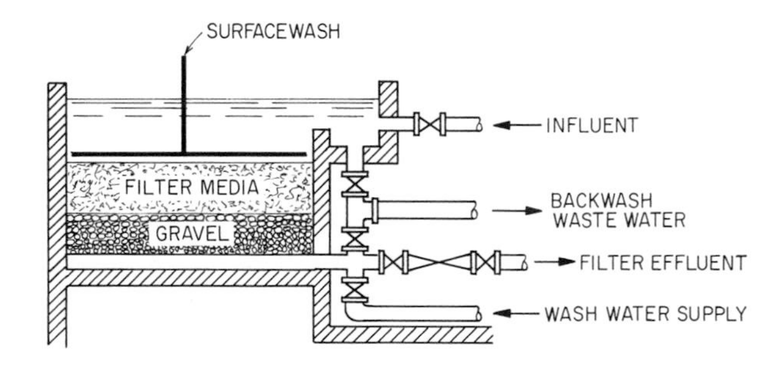

Figure 6–13 Cross section through a typical gravity filter.

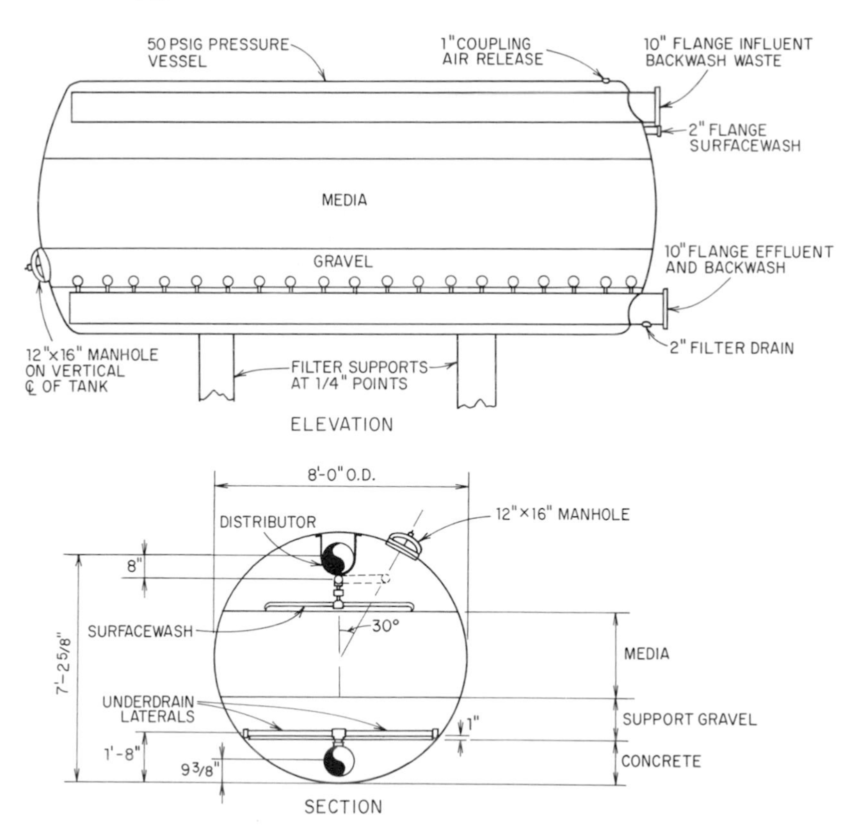

Figure 6–14 Typical pressure filter. *(Courtesy Neptune Microfloc, Inc.)*

Figure 6–15 Pressure filter installation at South Tahoe water reclamation plant. *(Reprinted from* Water & Wastes Engineering, *6:4, Apr., 1969, p. 36. R. H. Donnelley Corp.)*

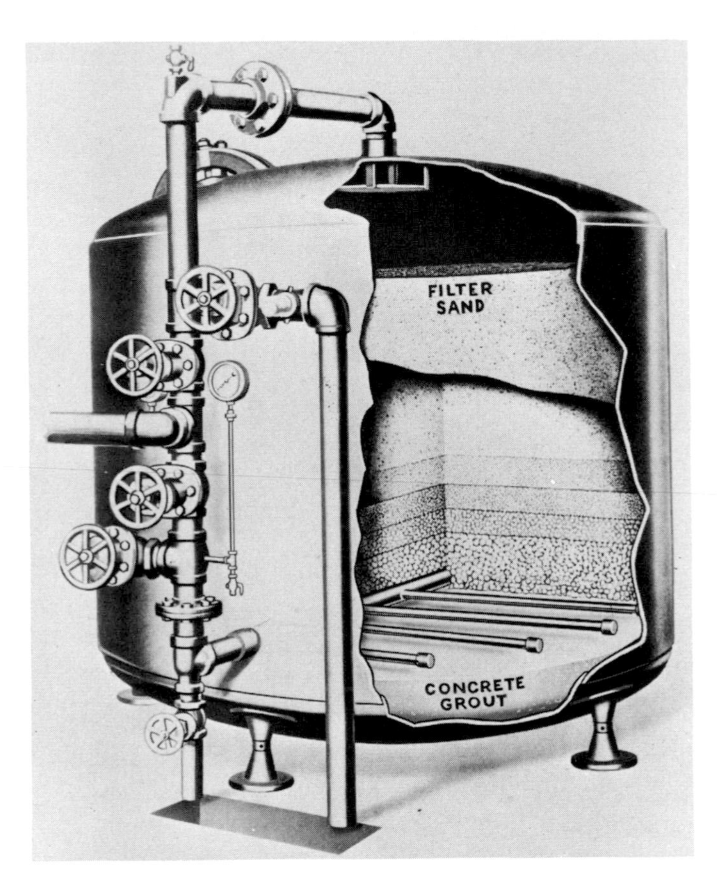

Figure 6–16 A typical vertical pressure filter with cement grout fill in the bottom head, pipe headers, and lateral underdrains, gravel supporting bed and filter sand. *(Courtesy Infilco Products)*

DESIGN AND OPERATION OF FILTER SYSTEMS

General Considerations

The typical pressure filtration plant shown in Figure 6-17 will be used to illustrate the design and operating considerations involved in planning a filter system. Some special considerations involved in gravity filter design will also be discussed.

It is not necessary to house the entire filter structure. Filter controls and pipe galleries should be housed. As can be noted in the photographs of the South Lake Tahoe filter installation (Figure 6-15) only the end of the pressure vessels connected to the filter piping is en-

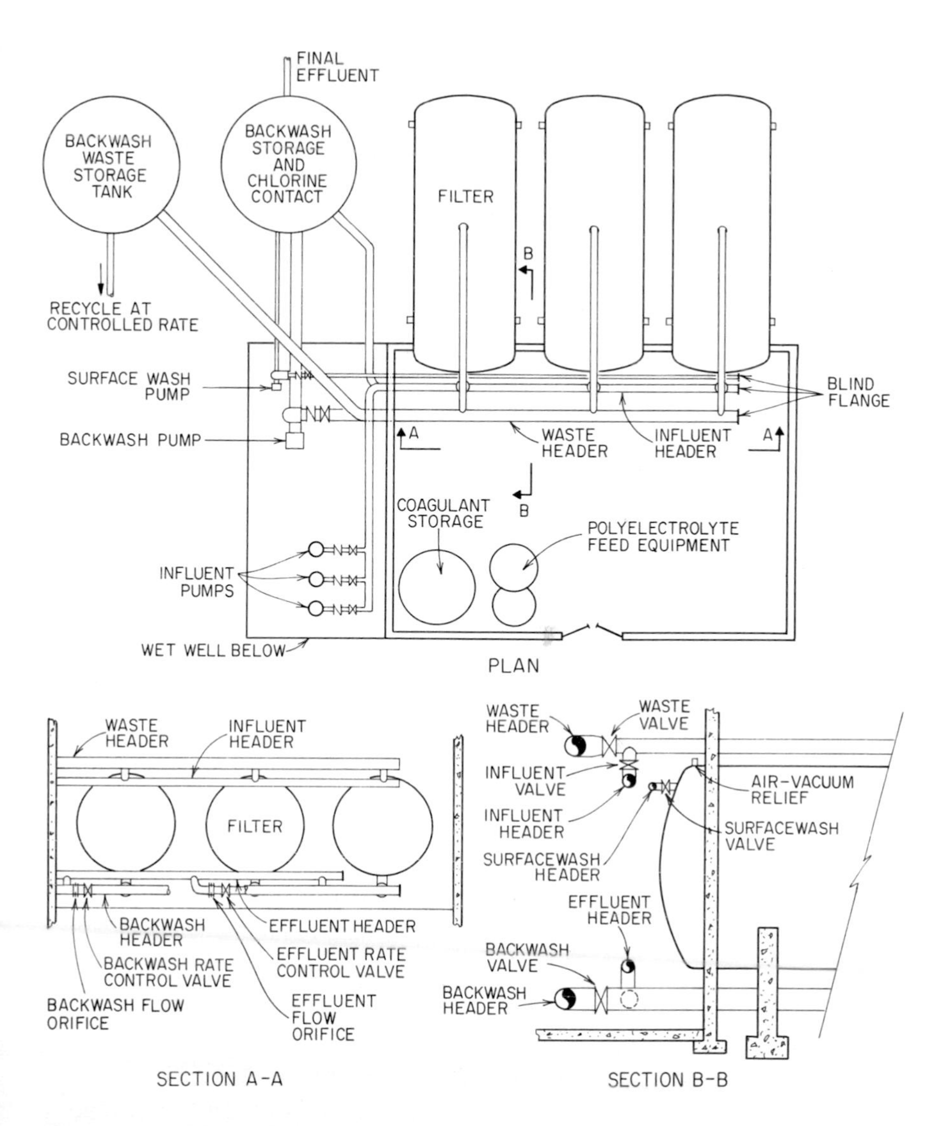

Figure 6–17 Typical layout of pressure filtration plant for wastewater treatment. *(Courtesy Neptune Microfloc, Inc.)*

closed. Despite the severely cold winters at this locale, no freezing problems have occurred. Of course, the filter vessels must be drained if they are removed from service during freezing weather. Gravity filters may be protected from freezing problems by placing a perforated pipeline around the perimeter of each filter just above the media surface through which air is bubbled to control ice formation.

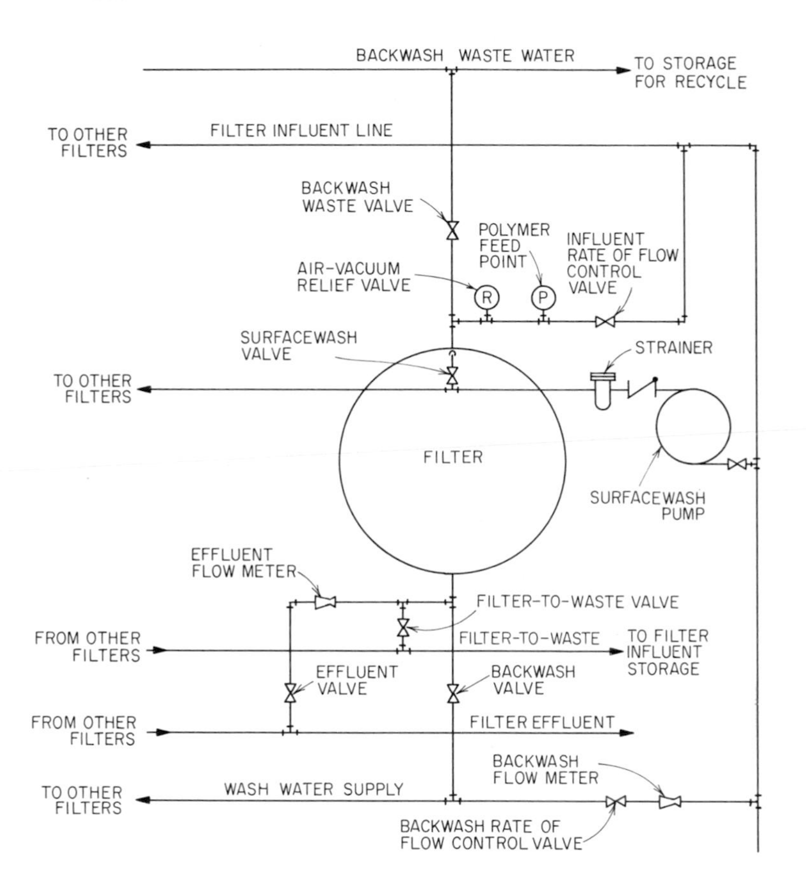

Figure 6–18 Illustrative pressure filter installation.

Figure 6-18 offers a schematic presentation of the piping and valving associated with the filter. The operation of the valves are summarized in Table 6-2.

The filter is equipped with a rate-of-flow controller to divide the flow among the filters, to limit the maximum flow, and to prevent sudden flow surges. A rubber-seated butterfly valve with an air-actuated hydraulic operator on the filter influent line is an acceptable means of flow control. The filter flow rate may be manually controlled by changing the valve position from a central control panel while observing the effluent flow indicator, mounted on the same panel. A valve stop may be used to control the degree of valve opening in the automatic cycle. By adjusting the stop, the valve will automatically open to set a predetermined initial flow rate. Of course, the filter rate will decrease as the filter head loss builds up. The

Table 6-2 Operational Cycle of Pressure Filter in Figure 6-18.

Valve	Filter Cycle	VALVE OPERATION Backwash Cycle	Filter-to-waste Cycle
Influent rate of flow control valve	open	closed	open
Effluent valve	open	closed	closed
Backwash rate of flow control valve	closed	open	closed
Backwash valve	closed	open	closed
Backwash waste valve	closed	open	closed
Surface wash valve	closed	open	closed
Filter to waste valve	closed	closed	open

amount of decrease will vary with the characteristics of the filter influent pump. Preferably, a constant rate of flow can be maintained by automatically positioning the influent valve with a signal from the filter flow meter. As the filter head loss increases and the flow begins to decrease, the influent valve is automatically opened further to maintain the current flow rate. There are a variety of commercially available filter flow control systems in addition to this illustrative one.

The filter run should be terminated when (1) the filter head loss reaches a predetermined value, (2) filter effluent turbidity exceeds the desired value, or (3) a predetermined amount of time passes. A differential pressure cell may be used to measure the head loss across the filter and can initiate a signal to start an automatic backwash program when the head loss reaches 15–20 ft. In addition to this function, measurement of filter headloss provides valuable information concerning plant performance. An increase in initial head loss over a period of time indicates clogging of the underdrain system or insufficient cleaning of the filters. The rate of head loss buildup during a run reflects the efficiency of pretreatment and aids in determining the optimum polymer feed dosage, as discussed previously.

Inexpensive, reliable turbidimeters are available to continuously monitor the quality of the filter effluent. Should the turbidity exceed a level suitable for a downstream process or for discharge, the signal from the turbidimeter can be used to initiate the backwash program or sound an alarm. Recording of the turbidimeter output provides a continuous record of filter performance. The Hach Low Range Turbidimeter, shown in Figure 6-19, a relatively inexpensive unit, is well suited to this application. It is a continuous-flow nephelometer. A strong beam of light passed through a water sample is scattered by particles of turbidity just as dust particles suspended in a room

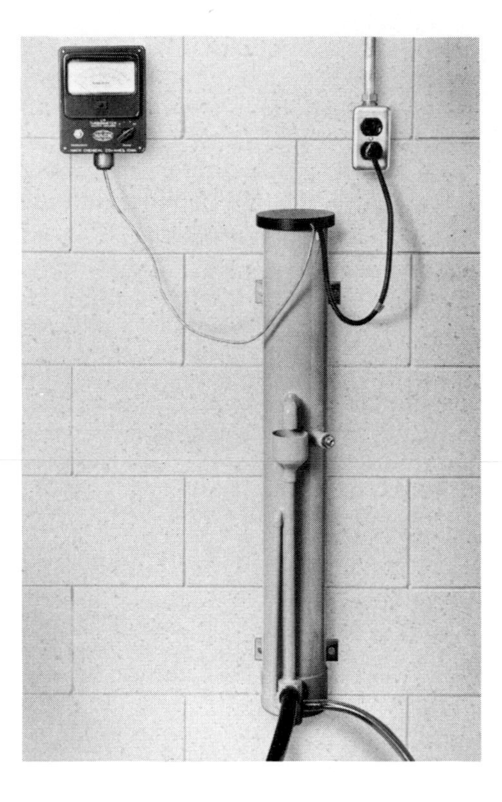

Figure 6–19 A continuous-flow turbidimeter capable of measuring turbidities of less than 0.1 JU. *(Courtesy Hach Chemical Co.)*

scatter sunlight streaming through a window. If the effluent is free of turbidity, no light is scattered and no light reaches the photoelectric light-measuring cells, resulting in a zero reading. The presence of turbidity in the sample results in light being scattered and some of the light falls on the photocells. The most important advantage of a nephelometer is that a very intense beam of light can be passed through the sample, and this results in very high sensitivity so that a trace amount of turbidity can be accurately measured. The output from the photocells is zero when the turbidity is zero and the output increases in proportion to the amount of turbidity read. The required sample stream volume is 0.25–0.50 gpm with a head of at least 6 in. for the Hach unit shown. The response time to detect a change in turbidity is 30 sec. The unit has ranges of 0–0.2 JU, 0–1.0 JU, 0–3.0 JU, and 0–30 JU. The approximate cost of a turbidimeter and recorder with alarm is $700 (1970 prices).

The backwash program may be carried out manually with local or remote controls or automatically. The automatic operation can be readily and economically provided with commercial sensing and control equipment and is strongly recommended.

At the end of a filter run, dependable equipment is available for fully automatic control of all valves, pumps, chemical feeders, and other process components as required to take the dirty filter off the line, backwash it at a controlled rate, provide filter-to-waste, and restore it to normal filtration service.

Some features will be described which can be incorporated into the automatic control of filter backwashing. The master control may be provided by an automatic sequencing circuit (step switch) which is interlocked so that the necessary prerequisites for each step are completed prior to proceeding to the next step. At the receipt of a backwash start signal, the following events will occur in the sequence listed in this illustrative program: Filter influent and effluent valves close. Any chemical feed stops to the filter being backwashed. Plant chemical feeds adjust to the new plant flow rate to maintain proper chemical feed to the filters still in service. The waste valve starts to open. When the waste valve reaches the fully open position and actuates a limit switch, the surface wash pump starts and the surface wash valves open. Surface wash flow to waste continues for a period of time adjustable up to 10 min. At the end of the initial surface wash period, usually 1–2 min, the main backwash valve opens. Backwash and surface wash both continue for a period of time, usually 6–7 min, adjustable up to 30 min. Backwash flow rate is indicated on a controller and is controlled automatically to a manual set point. At the end of the combined wash periods, the surface wash valves close and the surface wash pump stops. Backwash continues without surface wash for a period, usually 1–2 min, adjustable up to 30 min. At the completion of the backwash period, the backwash valve closes. After the surface wash valves and the backwash valve have closed, the waste valve starts to close. When the latter valve has closed, influent and effluent waste valves open and the bed filters to waste for a period of time, usually 3–7 min, adjustable up to 30 min. The backwash delay timer resets and begins a new timing cycle adjustable up to 12 hr. The bed selector switch steps to the next filter. Chemical feed to the clean filter is reestablished. At the end of the filter-to-waste period, the effluent waste valve closes and the effluent valve opens to restore the cleaned bed to normal filter service. Provision should be made for optional manual operation of all automatic features.

It may be desirable to alarm certain functions which affect filter operation on a conveniently located annunciator panel. These alarm functions include high turbidity, high head loss, low plant flow, low backwash flow rate, excessive length of backwash, and high head loss across surface wash water strainer.

Cast-iron pipe and fittings and coal-tar enamel-lined welded steel pipe and fittings are the most widely used materials for filter piping. The layout of filter piping must include consideration of ease of valve removal for repair and easy access for maintenance. Flexible pipe joints should be provided at all structure walls to prevent pipeline breaks due to differential settlement. The use of steel pipe can reduce flexible joint requirements. Color coding of the filter piping is a valuable operating aid. The filter piping is usually designed for the flows and velocities shown in Table 6-3. The rubber-seated, pneumatically actuated, hydraulically operated butterfly valve has almost entirely replaced the hydraulically actuated and operated gate valves that were formerly used extensively as filter valves. The butterfly valve is smaller, lighter, easier to install, better for throttling services, and can be installed in any position. The valves should be factory equipped with the desired valve stops, limit switches, and position indicators because field mounting of these devices is often unsatisfactory.

Table 6-3 Filter Piping Design Flows and Velocities.

Description	*Velocity (fps)*	*Maximum Flow per Filter Area Served (gpm/ft²)*
Influent	1–4	8–12
Effluent	3–6	8–12
Wash water supply	5–10	15–25
Backwash waste	3–8	15–25
Filter to waste	6–12	4–8

The concrete walls of gravity filters frequently must be designed to withstand hydrostatic pressure in two directions and must be designed for water tightness to prevent spalling when subjected to fluctuating water levels in cold climates. Minimum wall thicknesses of 9–12 in. should be used for water-bearing walls, with ample cover for reinforcing bars to protect them against corrosion should spalling occur.

Filter Underdrains

The primary purposes of the filter underdrain are to separate the water from the granular media through which it has passed and to distribute the filter and backwash flows evenly over the filter area. To achieve good distribution of backwash flow, a head loss of 3–15 feet is usually introduced across the underdrain. One of the most common underdrains is the perforated pipe lateral underdrain (Figure 6-16). This system uses a main header with several pipe laterals on both sides. The laterals have perforations on the underside so that the velocity of the jets from the perforations during backwash is dissipated against the filter floor and in the surrounding gravel. Piping materials most commonly used are steel, transite, or PVC. Usually, orifice diameters are ¼–½ in. with spacings of 3–8 in. Fair and Geyer (1958) present the following rules of thumb for pipe lateral underdrain design:

1. Ratio of area of orifice to area of bed served, 1.5×10^{-1}–5×10^{-3} : 1
2. Ratio of area of lateral to area of orifices served, 2 : 1–4 : 1
3. Ratio of area of main to area of laterals served, 1.5:1–3:1
4. Diameter of orifices, ¼–¾ in.
5. Spacing of orifices, 3–12 in. on centers
6. Spacing of laterals, closely approximating spacings of orifices

A graded gravel layer usually 14–18 in. deep is placed over the pipe lateral system to prevent the filter media from entering the lateral orifices and to aid in distribution of the backwash flow. A typical gravel design is shown in Table 6-4. A weakness of the gravel-pipe lateral system has been the tendency for the gravel to eventually intermix with the filter media. These gravel "upsets" are caused by localized high velocity during backwash, introduction of air into the backwash system, or use of excessive backwash flow rates. The gravel layer can be stabilized by using 3 in. of 1-mm garnet or ilmenite as

Table 6-4 Typical Gravel Bed for Pipe Underdrain System.

Description	*Number of Layer*				
	1	2	3	4	5
Depth of layer (inches)	[1]	3	3	4	4
Sq. mesh screen opening (inches)					
Passing	1	3/4	1/2	1/4	1/8
Retained	3/4	1/2	1/4	1/8	1/16

[1]Bottom layer should extend to a point 4 in. above the highest outlet of wash water.

the top layer of the gravel bed. This coarse, very heavy material will not fluidize during backwash and provides excellent stabilization for the gravel. It also prevents the fine garnet or ilmenite used in a mixed, tri-media filter from mixing with the gravel support bed. The remaining major disadvantage of the gravel-pipe lateral system is the vertical space required for the gravel bed. Gravel layers are also used with several of the commercially available underdrain systems such as the Leopold Bottoms (see Figures 6-20 and 6-21), Wheeler Bottoms, and Wagner Bottoms. Gravel depths and gradations vary for these underdrain systems. For example, Leopold recommends the following

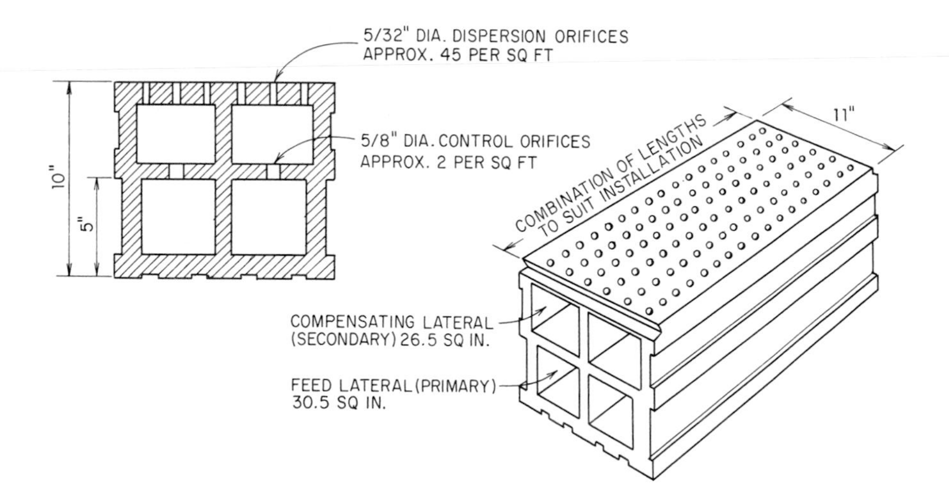

Figure 6–20 Leopold filter bottom. *(Courtesy F. B. Leopold Co.)*

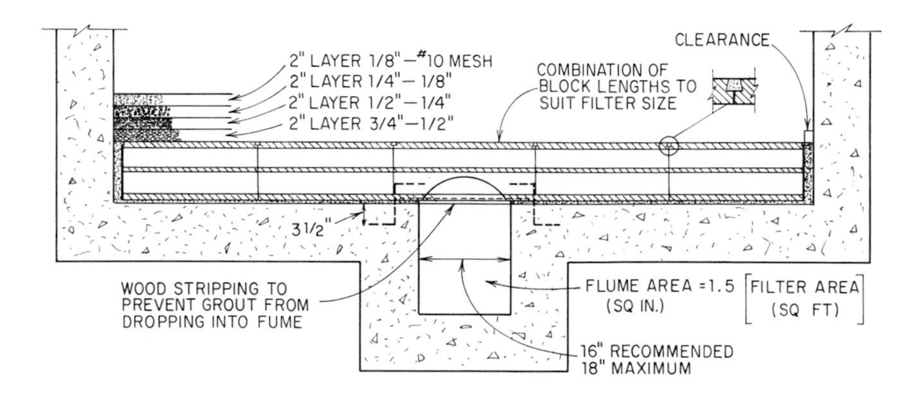

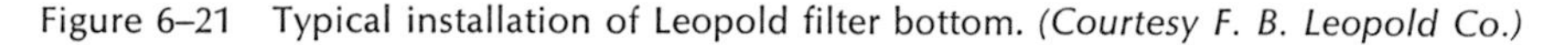

Figure 6–21 Typical installation of Leopold filter bottom. *(Courtesy F. B. Leopold Co.)*

Figure 6–22 The Camp filter underdrain nozzle. *(Courtesy Walker Process Equipment, Inc.)*

gradation: 2–¾ × ½ in.; 2–½ × ¼ in.; 2–¼ × ⅛ in.; 2–⅛ × 10 mesh.

Several underdrain systems have been developed with the goal of eliminating the need for the gravel support bed. These systems employ strainerlike nozzles or porous plates. Plastic or steel nozzles of various shapes are available from several manufacturers. Among these are the Eimco Corporation, which manufactures a plastic, conical nozzle with a screen supported between two layers of plastic. Degremont Company makes plastic nozzles which have finely slotted sides. The Edward Johnson Company manufactures a stainless steel nozzle consisting of 4¾-in.-diameter well screen pipe with a cap welded on one end. Walker process offers a nozzle made of thermoplastic resin having 0.25-mm aperature vertical slots (Figure 6-22). The nozzles are mounted in hollow, vitrified clay blocks laid to form the filter bottom. The 9-in.-high hollow-core blocks are grouted in place and provide a system of ducts for filtered water flow as well as for backwash flow. The nozzles are designed with built-in orifices to meter the backwash flow evenly over the filter area. Care must be taken in selection of these nozzles to insure that the finest filter media will not pass through the nozzle openings. The nozzles are mounted either in a false bottom or in a pipe lateral. The disadvantages of the plastic nozzles are that they are fairly fragile and can be easily broken when being installed or when the medium is placed. The chief operating problem noted by the authors in use of the nozzles is plugging from the inside with material found in the backwash water or with rust or other particulate matter from the filter backwash piping. This plugging frequently occurs on the first application of backwash due to the failure to adequately clean the construction debris from the system.

Porous plates made of aluminum oxide, such as manufactured by the Carborundum Company, or of stainless steel, such as manufactured by the Multi-Metal Wire Cloth Company, are also available to eliminate the need for the gravel layer. The aluminum oxide plates are usually mounted on steel or concrete frames which support the plate on at least two edges, or by studs projecting from the filter floor fastened to the corners of the plates. Mastic or caulking compounds are used to seal the joints. The disadvantages of the aluminum oxide plates are that they are brittle and easily broken during installation; the sealing of joints between plates may be difficult; and the plates can become plugged from either side. If plugged plates are not promptly cleaned, structural failure may occur due to decreasing differential pressures across the plates during backwash. They offer the advantages of good flow distribution and minimum filter height since the media can be placed directly on the plates. The metallic plates (trade name, Neva-Clog) are made up of two steel plates perforated by round holes with the holes offset such that perforations of one sheet face the unperforated portion of the other. The sheets are spot-welded together with very small spacing between the sheets. Fluids pass through the holes of the top sheet, turn at right angles to pass through the narrow space between the sheets, and then turn again at right angles to pass through holes in the bottom sheet. Solids larger than 100 mesh are too large to pass through and are retained. These metal plates have an advantage over the aluminum oxide plates in that the only sealing required is where the plate meets the filter walls. However, the steel plates must be supported by a steel or fiberglass grid.

It is apparent from this discussion that all of the available underdrain systems have strengths and weaknesses and must be evaluated in light of each specific application.

Backwashing of Filters

Effective cleaning of the filter media during backwash is essential to successful plant operation. For many years it was felt that a 50-percent expansion of the filter media was required for effective cleaning. However, it is recognized that optimum scouring of the particles results when the media are just suspended. When suspended, the shearing force past the grains of media is equal to the weight in water and is not increased with further expansion and higher wash rates. Camp (1961) reports that a wash rate corresponding to 25-percent expansion of the coal layer is adequate for a dual media filter but that a wash rate corresponding to 31 percent coal expansion provides a safety factor. The wash rate required to achieve a given expansion depends upon the water temperature and the size and specific gravity of the filter media. A rise in water temperature from 10–20°C will necessitate an increase in wash rate of about 30 percent to maintain a given expansion of a silica sand. Of course, the required wash rate increases with increasing specific gravity for a given size particle and with increasing particle size for a given specific gravity. For example, an increase in silica sand size from 0.35–0.63 mm requires an increase in wash rate from 12.5–29 gpm/ft^2 for 30 percent expansion at 18°C. A mixed-media design typically used for removal of moderate quantities of chemical floc (3 in. of $-40+80$ garnet or ilmenite, 9 in. of $-20+40$ silica sand, and 18 in. of -10×20 coal) requires a backwash rate of about 15 gpm/ft^2 at 18°C. A coarser mixed-media filter such as used for removal of biological floc may require a backwash rate of 18–20 gpm/ft^2, depending upon the specific media selected. The head loss through the expanded filter is 2–4 ft, which is relatively small compared to the head loss through the filter underdrain system that is induced to distribute the wash water.

The required duration of backwash depends on the particular service but it is typically 5–8 min. The source of backwash water in sewage filtration applications should be filter effluent or chemically coagulated and settled effluent rather than secondary effluent to insure that the backwash supply will always be free of large quantities of suspended solids. A backwash flow indicator should be included so that the operator can be sure that the desired backwash rate is being maintained at all times.

As noted earlier, the introduction of air into a gravel support bed can overturn the gravel and disrupt filter operation. Air can be unintentionally introduced to the bottom of the filter in a number of ways. If a vertical pump is used for backwash, air may collect in the vertical pump column between backwashings. This air can be eliminated by starting the backwash pump against a closed backwash valve and releasing the air through a pressure release valve in the backwash line. Oxygen may be released from the water standing in the backwash piping and be carried into the filter by the wash water. This air can be released by placing a pressure release valve at the high point in the wash water line with a pressure water connection to that line to keep it full of water and expel the air.

Care must be taken to design the entry of the wash water into the filter bottom to dissipate the velocity head so that uniform distribution of wash water is obtained.

The backwash wastewater must be reprocessed. The rate of backwash flow, if returned directly to an upstream clarifier, is usually large enough in relation to the design flow through the clarifier to cause a hydraulic overload and upset of the clarifier. In this case, the backwash wastes should be collected in a storage tank and recycled at a controlled rate. The volume of backwash wastewater is typically 2–5 percent of the plant throughput and plant components must be sized to handle this recycled flow. Provisions must be made to store the incoming flow during the backwash cycle or, if there are parallel units, to increase the rate on the other filters during the backwash cycle.

Effective surface wash is a necessity to insure that clumps of media and floc do not survive the backwash as "mud balls." Actually, the term "surface wash" is a misnomer when applied to efficient surface wash equipment because the lateral currents set up by the surface wash circulate the entire bed contents. A typical surface wash apparatus is represented by Palmer surface agitators (marketed by F. B. Leopold Company). This equipment consists of a revolving pipe with a number of nozzles suspended at its center by a bearing. These units are positioned just slightly (1–2 in.) above the normal surface of the filter. Wash water under high pressure (50–100 psi) is injected through the nozzles causing the pipe to rotate during the backwash cycle. Water supply requirements for these rotating surface wash devices varies but is generally in the range of 0.75–1.0 gpm/ft^2. Normally, surface wash is initiated about 1 min before the main backwash pumps are turned on and is stopped about 1 min prior to the end of backwash to permit the mixed-media bed to reclassify. When in operation, the surface wash circulates the entire bed contents. The nozzles are directed downward about 30 deg from the horizontal and will cause agitation throughout the bed depth rather than just at the surface. The source of water used for the surface wash must be of high quality to insure that plugging of the nozzles does not occur. A strainer with 3/32-in. perforations in the surface wash line is a wise precaution. In pressure fllter systems, an external indicator (available from Neptune Microfloc) should be used to show that the surface wash arm is actually rotating during backwash. A rotating agitator similar to the surface wash arms may be installed within the bed in very deep filters. These subsurface agitators have the nozzle discharge staggered 15 deg up to 15 deg down.

An alternate to mechanical surface wash employment is the use of airwater backwash techniques in which air is injected through the filter underdrain system to break up mud balls as it rises through the filter. These techniques have been used fairly extensively in Europe but lost favor in the United States about 30 years ago due to problems encountered with upsetting of beds and loss of media. Recent improvements in underdrain systems have led to renewed interest. Because of the tendency of the air to float coal out of a filter, the backwash water and the air cannot be applied at the same time with dual or tri-media beds utilizing coal. The air-water technique applied to these beds usually consists of the following steps: (1) stop the influent and lower the water level to within 2–6 in. of the filter surface; (2) apply air alone at a rate 2–5 cfm/ft^2 for 3–10 min; (3) apply a small amount of backwash water (about 2 gpm/ft^2) with the air continuing until the water is within 8–12 in. of the wash water trough (the air is then shut off); (4) continue water backwash at 8–10 gpm/ft^2 until the filter is clean.

An important point to consider is that a mixed-media bed will require 1–2 min of wash at 12–15 gpm/ft^2 at the end of the wash period to properly classify the media.

It has been the authors' experience that although lower backwash rates are used with the air-water techniques, the amount of backwash water required to achieve the *same degree of cleaning* is not significantly different than with surface wash in conjunction with backwash. The air-water techniques increase the complexity of the backwash cycle, lengthen the filter downtime for backwashing, increase the risk of media loss, and offer little, if any, advantage in the quantity of backwash water required to achieve a given degree of cleaning.

MIXED-MEDIA FILTRATION AT SOUTH TAHOE

The key to the reliable performance and continuous operation of the advanced wastewater treatment plant at South Tahoe is the successful development of the mixed media filter. It is a unit process which is vital to the proper functioning of the plant as a whole. The beds will accept heavy shock loads of suspended solids from upsets in biological or chemical pretreatment without interruption in service or deterioration in effluent quality. They remove all suspended solids, and produce excellent water clarity (turbidity = 0.01–1.0 JU). They remove significant amounts of colloidal and dissolved phosphorus from the wastewater. They protect the granular carbon treatment which follows from prolonged interruptions in service, and serious loss of efficiency which would occur in the absence of the filters due to poor applied water quality.

For the design flow of 7.5 mgd, there are 3 pairs of pressure beds in series, each 10 ft in diameter by 38 ft long. The design filter rate is 5 gpm/ft^2, but rates as high as 8 gpm/ft^2 have been employed at full treatment efficiency. The backwash rate is 15 gpm/ft^2. Each pair of beds is washed as a unit in series. The surface wash consists of four 7-ft-diameter rotary filter agitators per bed. Each bed consists of 3 ft of mixed media (as supplied by Neptune Microfloc), supported on 3 in. of coarse garnet and 2 ft, 4 in. of graded gravel. The underdrains are perforated plastic pipe. The influent rate-of-flow controller consists of a Dall flow tube and a rubber-seated butterfly valve. Loss of head across each bed is continuously measured and recorded. Turbidity of separation bed effluent is continuously measured by a Hach CR Turbidimeter to tenths of a Jackson unit and recorded. All filter operations are fully automatic. Backwash is initiated by time clock, high head loss, high turbidity, or manually. The beds are backwashed, filtered to waste, and restored on line automatically by a program timer. The filters are backwashed with filter influent water by means of a pump. There is a pressure booster pump for surface wash supply. Waste wash water discharges into an 80,000 gal steel tank (which holds the water from two backwashes.) The water from this tank is returned to the treatment process slowly over a period of about 2 hr. One pair of these beds has been in service for 5 years, and the other two pairs have been in use for 2 years at this writing (1970).

All except one end of each bed is installed out-of-doors. An allowance was made in the design for the formation of 4 in. of ice inside the steel filter shell. This ice then insulates the tank against further freezing under conditions of normal water flow through the bed.

The performance of these beds and the control system has been excellent. The length of filter runs varies from 4 hr under very bad conditions to about 60 hr under good. They have been used with alum as the primary coagulant in pretreatment, and with Calgon ST-270 or Purifloc N-11 (0.1–0.8 mg/l) as a filter aid. They have also been used with lime as the primary coagulant in pretreatment and either alum (1–20 mg/l) or ST-270 or N-11 (0.01–0.10 mg/l) as a filter aid applied directly to the filter influent. Normally the beds are backwashed when the head loss through each bed is about 8 ft (16 ft total). However, they have been backwashed successfully after head losses through the two beds totaled as much as 40 ft. This high head loss would be excessive for continuous operation. Table 6-5 indicates typical removals of several materials by the separation beds.

Table 6-5 Typical Removals by Tahoe Mixed-Media Filters.

Substance	*Typical Concentrations (mg/l)* *Influent*	*Effluent*	*Range (% removal)*
Phosphorus, total	0.65	0.05	70–95
Phosphorus, dissolved	0.45	0.05	65–90
Phosphorus, particulate	0.20	0.00	100
COD	23	15	20–45
BOD	9	5	40–70
SS	15	0	100
Turbidity, JU	7.0	0.3	60–95

The capital costs for the 7.5 mgd filtration system including the separation beds, media, automatic control system, all auxiliary equipment, and building is $687,000 (at the 1969 FWPCA STP Construction Cost Index = 127.1). The total costs for operation and maintenance at design flow is about $24/mg including amortization.

References

1. Anonymous, "Renovation and Reuse of Wastewaters in Britain." Published by Water Pollution Research Laboratory, 1967.
2. Bodien, D. G., and Stenburg, R. L., "Microscreening Effectively Polishes Activated Sludge Plant Effluent," *Water & Wastes Engineering*, Sept., 1966, p. 74.
3. Boucher, P. L., "A New Measure of the Filterability of Fluids with Application to Water Engineering," *ICE Journal*, 1947, p. 415.
4. Camp, T. R., "Discussion of Paper by W. R. Conley," *Journal American Water Works Assoc.*, 1961, p. 1478.
5. Conley, W. R., Jr., "Integration of the Clarification Process," *Journal American Water Works Assoc.*, 1965, p. 1333.
6. ———, and Hsiung, K., "Design and Application of Multimedia Filters," *Journal American Water Works Assoc.*, 1969, p. 97.
7. Convery, J. J., "Solids Removal Processes," FWPCA Symposium on Nutrient Removal and Advanced Waste Treatment, Tampa, Florida, Nov., 1968.
8. Culp, G. L., "Secondary Plant Effluent Polishing," *Water and Sewage Works*, Apr., 1968, p. 145.
9. ———, and Hansen, S. P., "Extended Aeration Effluent Polishing by Mixed Media Filtration," *Water and Sewage Works*, Feb., 1967, p. 46.

10. ———, "How to Reclaim Wastewater for Reuse," *American City,* June, 1967, p. 96.
11. Diaper, E. W. J., "Microstraining and Ozonation of Sewage Effluents." Presented at the Forty-first Annual Conference of the Water Pollution Control Federation, Chicago, Illinois (Sept., 1968).
12. ———, "Tertiary Treatment by Microstraining," *Water and Sewage Works,* June, 1969, p. 202.
13. Dixon, R. M., and Evans, G. R., "Experiences with Microstraining on Trickling Filter Effluents in Texas." Presented at the Forty-eighth Texas Water and Sewage Works Associations Short School, College Station, Texas (March, 1966).
14. Eliassen, R. and Bennett, G. E., "Renovation of Domestic and Industrial Waste Water." Presented at the International Conference on Water for Peace, Washington, D.C. (May, 1967).
15. Evans, S. C., "Ten Years Operation and Development at Luton Sewage Treatment Works," *Water and Sewage Works,* 1957, p. 214.
16. Fair, G. M., and Geyer, J. C., *Elements of Water Supply and Waste-Water Disposal,* John Wiley and Sons, New York, 1958.
17. *Filtration, Water Treatment Plant Design Manual,* American Water Works Assoc., 2 Park Ave., New York, 1969.
18. Lynam, B., Ettelt, G., and McAloon, T., "Tertiary Treatment at Metro Chicago By Means of Rapid Sand Filtration and Microstrainers," *Journal Water Pollution Control Federation,* 1969, p. 247.
19. New England Interstate Water Pollution Control Commission, "A Study of Small, Complete Mixing, Extended Aeration Activated Sludge Plants in Massachusetts" 1961.
20. Oakley, H. R., Personal Communication (Jan. 7, 1969).
21. ———, and Cripps, T., "British Practice in the Tertiary Treatment of Sewage," *Journal Water Pollution Control Federation,* 1969, p. 36.
22. O'Farrell, T. P., Bishop, D. F., and Bennett, S. M., "Advanced Waste Treatment at Washington, D.C." Presented at the Sixty-fifth Annual AICHE meeting, Cleveland, Ohio (May, 1969).
23. Shatto, J., "Aerobic Digestion and Diatomite Filter," *Public Works,* Dec., 1960, p. 82.
24. Smith, R., "Cost of Conventional and Advanced Treatment of Wastewaters," *Journal Water Pollution Control Federation,* 1968, p. 1546.
25. Tchobanoglous, G., "Filtration Techniques in Tertiary Treatment." Presented at the Fortieth Annual Conference of the California Water Pollution Control Association (Apr. 26, 1968).
26. ———, and Eliassen, R., "Filtration of Treated Sewage Effluent," *Journal Sanitary Engineering Division,* American Society of Civil Engineers, 1970, p. 243.
27. ———, and Bennett, G., "Progress Report, Water Reclamation Study Program." FWPCA Demonstration Grant, WPD 21–05 (Oct., 1967).
28. Truesdale, G. A., and Birkbeck, A. E., "Tertiary Treatment Processes for Sewage Works Effluents." Presented to the Scottish Centre of the Institution of Public Health Engineers (Mar. 28, 1966).

7

Granular Activated Carbon

PURPOSE

The use of granular activated carbon for the adsorption of organic materials from wastewater has become firmly established as a practical, reliable, and economical unit process for water pollution control. It is a valuable new tool for which sanitary engineers will have many uses and will make wide application in preserving water quality, meeting discharge requirements, and producing reclaimed water free of color, odor, froth, and other evidences of the presence of organic pollutants. Historically, in wastewater treatment, some organic materials have been removed by biological oxidation methods, including activitated sludge treatment and trickling filtration. These conventional processes may remove nearly all of those organics measured by the biochemical oxygen demand (BOD) test, but are ineffective in removing the so-called refractory organic materials as measured by the chemical oxygen demand (COD) test. Even well-treated secondary effluents contain 50–120 mg/l of organics. These materials include tannins, lignins, ethers, proteinaceous substances, and other color and odor producing organics, as well as MBAS (methylene blue active substances), herbicides, and pesticides such as DDT. Certain refractory organic substances added to the water in a stream may contribute to algal growth, contribute to fish kills and tainting fish flesh, produce taste and odor in water supplies withdrawn from the stream, and may have cumulative harmful physiological effects if present in drinking water.

There are a rather limited number of unit processes which are capable of removing these refractory organic materials from wastewater, including reverse osmosis, freezing, chemical oxidation, distillation, adsorption on powdered carbon, and adsorption on granular carbon. Powered activated carbon could be used on a once-through basis but there are severe dust problems in handling the rather large quantities needed in wastewater treatment. There is also a problem of disposal of the spent carbon unless it is incinerated along with the sewage sludge. Powdered carbon used in sugar refineries was reactivated with reasonable success for many years but its use was discontinued in favor of the use of granular carbon. This change came about as the result of problems in regeneration of powdered carbon with dust and dirt, the production of carbon fines so small that they were not filterable, and poor recoveries ranging from 60–90 percent. It is possible that some of the current research which is under way in regenerating powdered carbon will find ways to correct these deficiencies (see Chapter 10) but the solutions do not appear to be at hand now. Reverse osmosis, freezing, and distillation are not yet developed to the point of being either practical or economical processes for removing organics from water. Adsorption on granular activated carbon is at this time the best method for this purpose. The commercial availability of a high-activity, hard, dense granular activated carbon made from coal, plus the development of multiple hearth furnaces for on-site regeneration of this type of carbon have drastically reduced the cost of granular activated carbon as a unit process for wastewater treatment. Beds of granular carbon have the adsorptive capacity to handle shock hydraulic or organic loads with no loss in efficiency.

Presently, the major manufacturers of granular activated carbon include Calgon Corp. (formerly Pittsburg Activated Carbon Co.), West Virginia Pulp and Paper Co., Atlas Chemical Industries, Inc., National Carbon Co., American Norit Co., Inc., and Witco Chemical Co., Inc.

HOW CARBON ADSORBS ORGANICS

Upon contact with a water containing soluble organic materials, granular activated carbon selectively removes these materials by adsorption. Adsorption is the phenomenon whereby molecules adhere to a surface with which they come into contact, due to forces of attraction at the surface. The use of surface energy to attract and hold molecules is physical adsorption. The fact that activated carbon has an extremely large surface area per unit weight (on the order of 1,000 m^2/g) makes it an extremely efficient adsorptive material. The activation of carbon in its manufacture produces many pores within the particles, and it is the vast areas of the walls within these pores that accounts for most of the total surface area of the carbon. In water, activated carbon has a preference for large organic molecules and for substances which are nonpolar in nature. The forces of attraction between the carbon and the adsorbed molecules are greater the closer the molecules are in size to the pores. The best adsorption takes place when the pores are just large enough to admit the molecules.

Crushing of carbon particles to produce smaller particles enhances the rate of adsorption by exposing more entrances to the carbon pores. Because the carbon particle size primarily affects the rate of adsorption and not the total adsorptive capacity of the carbon, the difference in performance of columns containing different size carbons decreases as the contact time increases.

The rate of adsorption of MBAS and other organics found in wastewater increases with decreasing *p*H of the water. Adsorption is very poor at *p*H values above 9.0.

The effect of turbidity or suspended solids in water applied to granular carbon on the efficiency and life of the carbon has not been determined precisely. However, it is evident that any restriction of pore openings or buildup of ash or other materials within the pore openings due to the presence of suspended or colloidal materials and their accumulation on or in the carbon particles might have an adverse effect upon the adsorptive capacity or service life of the carbon. These hazards can be minimized by applying water which has been pretreated to the highest practical clarity to the carbon.

CARBON SPECIFICATIONS

Presently the best carbons for treatment of wastewater appear to be those made from select grades of coal. These carbons are hard and dense and can be conveyed in water slurry with no appreciable deterioration. The physical strength of the carbon must be great enough to withstand the repeated handling required during regeneration.

The two most popular sizes of granular carbon for wastewater treatment are nominally 8 × 30 mesh and 12 × 40 mesh. The finer material has a higher rate of adsorption, but also has a higher head loss per unit depth of bed, and since the beds have lower porosity, they have a greater tendency to plug with materials filtered out of the wastewater. The 8 × 30 mesh carbon is the better choice. It reduces losses during regeneration and greatly simplifies carbon column operation, at only slight loss in efficiency. Recommend carbon specifications are given in Table 7-1.

Table 7-1 Suggested Specifications for Granular Activated for Use in Wastewater Treatment.

Total surface area (m^2/g)[1]	950–1500
Bulk density (lb/ft^3)	26
Particle density, wetted in water (g/cc)	1.3–1.4
Effective size (mm)	0.8–0.9
Uniformity coefficient	1.9 or less
U.S. Standard Series, sieve size	
larger than No. 8	max. 8%
smaller than No. 30	max. 5%
Mean particle diameter (mm)	1.5–1.7
Iodine number	min. 900
Abrasion number	min. 70
Ash	max. 8%
Moisture	max. 2%

[1]Brunauer, Emmett, and Teller (BET) method, *Journal of the American Chemical Society,* 60:309 (1938).

In the upflow countercurrent carbon columns at Tahoe, the capacity of 8 × 30 mesh carbon for COD is 0.5 lb of COD per pound of carbon.

Granular carbon is usually packaged in four-ply polyethylene reinforced kraft bags, 61 lb gross. They are available on trailer-truck pallets, 42 × 54 in., 38 bags per pallet, with a minimum of 16 pallets accepted on a 40-ft trailer. Railcar pallets are 48 × 48 in., 42 bags per pallet, with 24 pallets maximum per standard 50-ft railcar. Minimum bulk shipments are 30,000 lb by truck and 50,000 lb by rail.

SELECTING ACTIVATED CARBON FOR WASTEWATER TREATMENT

The selection of the carbon specified above appears on the basis of present knowledge to be the best that can be made for treatment of wastewater from those granular carbons now available. It must be pointed out however, that information and methods now available for selection are not fully developed and some of the decisions involved are necessarily somewhat arbitrary. For liquid adsorption systems there is no precise method for predicting the performance of carbons founded on their basic properties or those of the adsorbing molecules. In wastewater, the substances to be removed such as color, odor, COD, MBAS, and other refractory organics are always a composite of ingredients of unknown identity.

Further, the use of granular carbon is fairly new for treatment of wastewater, the potential market is great, and the competition among carbon manufacturers is likely to be keen. Therefore, it is not unreasonable to expect not only a future reduction in prices for granular carbon, but also a possible improvement in the kind and quality of carbons commercially available.

Sanitary engineers, chemists, and others responsible for selecting or purchasing these materials should be aware of the best methods for evaluating and characterizing carbons. These methods include quantifying the adsorptive characteristics, describing the physical properties, and conducting and interpreting pilot carbon column studies. With the use of these techniques, advantage can be taken of new developments and improvements in the field as they occur.

The adsorptive capacity of a carbon can be measured to a fair degree by determining the adsorption isotherm experimentally in the system under consideration. Simpler capacity tests such as the Iodine Number or the Molasses Number also may be an appropriate measure of adsorptive capacity.

The adsorption isotherm is the relationship, at a given temperature, between the amount of a substance adsorbed and its concentration in the surrounding solution. If a color adsorption isotherm is taken as an example, the adsorption isotherm would consist of a curve plotted with residual color in the water as the abscissa, and the color adsorbed per gram of carbon as the ordinate. A reading taken at any point on the isotherm gives the amount of color adsorbed per unit weight of carbon, which is the carbon adsorptive capacity at a particular color concentration and water temperature. In very dilute solutions, such as wastewater, a logarithmic isotherm plotting usually gives a straight line. In this connection, a useful formula is the Freundlich equation, which relates the amount of impurity in the solution to that adsorbed as follows:

$$x/m = KC^{1/n}$$

where x = amount of color adsorbed
m = weight of carbon
x/m = amount of color adsorbed per unit weight of carbon
k and n are constants
C = unadsorbed concentration of color left in solution

In logarithmic form:

$$\log x/m = \log k + 1/n \log C$$

in which $1/n$ represents the slope of the straight line isotherm.

Detailed procedures for establishing the experimental conditions and conducting isotherm adsorption tests are presented later in this book in Chapter 11. Test results for x/m against C are usually plotted on log-log paper to obtain the adsorption isotherm.

From an isotherm test it can be determined whether or not a particular purification can be effected. It will also show the approximate capacity of the carbon for the application, and provide a rough estimate of the carbon dosage required. Isotherm tests also afford a convenient means of studying the effects of *p*H and temperature on adsorption. Isotherms put a large amount of data into concise form for ready evaluation and interpretation. Isotherms obtained under identical conditions using the same test solutions for two test carbons can be quickly and conveniently compared to reveal the relative merits of the carbons.

Figures 7–1 and 7–2 are presented to illustrate the interpretation of adsorption isotherms. In Figure 7–1 the isotherm for carbon A is at a high level and has only a slight slope. This means that adsorption is large over the entire range of concentrations studied. The fact that the isotherm for carbon B in Figure 7–1 is at a lower level indicates

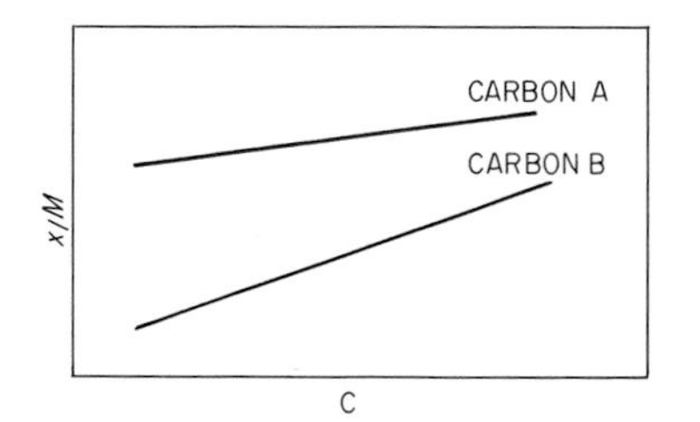

Figure 7–1 Adsorption isotherm, carbon A and B.

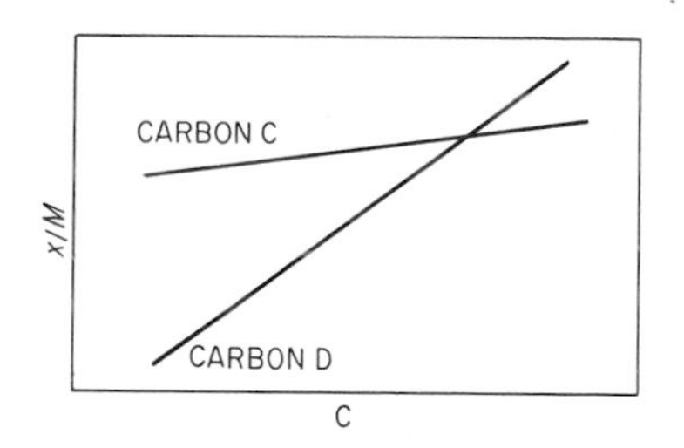

Figure 7–2 Adsorption isotherm, carbon C and D.

proportionally less adsorption, although adsorption improves at higher concentrations over that at low concentrations. An isotherm having a steep slope indicates that adsorption is good at high concentrations but much less at low concentration. In general, the steeper the slope of its isotherm the greater the efficiency of a carbon in column operation. In Figure 7–2, carbon D is better suited to countercurrent column operation than carbon C. It has a higher capacity at the influent concentration, or more reserve capacity. Carbon C in Figure 7–2 would be better than carbon D for batch treatment.

As mentioned previously, the Iodine Number and the Molasses Number also give an indication of the adsorptive capacity of a carbon. The Iodine Number is the milligrams of iodine adsorbed from a 0.02 *N* solution at equilibrium under specified conditions. The Molasses Number is an index of the adsorptive capacity of the carbon for color bodies in a standard molasses solution as compared to a standard carbon. The procedures for determining the Iodine and Molasses Numbers are given later in this book (Chapter 11). These tests are generally used for screening purposes. The Iodine Number gives a general indication of the efficiency of the carbon in adsorbing small molecules, and the Molasses Number the same for large molecules.

The physical properties of the carbon also are a clue to its probable performance. Physical factors include total surface area, particle size distribution, density, and abrasion number.

In plant-scale use of granular activated carbon, water is passed through a bed of carbon. Since this is a dynamic system, not only the equilibrium adsorption properties of the carbon are important, but the rates of adsorption as well. The flow rate and bed depth necessary for optimum performance will depend upon the rate at which impurities are adsorbed by the carbon. These determinations can best be made by dynamic pilot column tests. The general range of these variables to be investigated are flow rates of 2–10 gpm/ft^2 of column cross-sectional area, and bed depths of 10–30 ft. The diameter of the

Figure 7–3 Upflow pilot carbon column. *(Courtesy Cornell, Howland, Hayes & Merryfield)*

carbon columns can be scaled down to 2–8 in. by appropriate reductions in total flow, and still simulate full-scale plant conditions. If the laboratory cannot accommodate full height columns, the same total bed depth can be provided by several shorter columns operated in series. Figure 7–3 illustrates a pilot column arrangement used at the South Tahoe Water Reclamation Plant.

PILOT CARBON COLUMN TESTS

Pilot plant carbon column tests may be performed for the express purpose of obtaining design data for full-scale plant construction, rather than for research or other purposes.

What information from pilot plant studies is needed for plant design? It is assumed that one or two carbons have been selected which are effective in treating the particular wastewater, since this can be done in the laboratory by adsorption isotherms and Iodine Numbers, as already discussed. It is not necessary to run pilot column tests to reach this point. Pilot column tests make it possible to do the following things:

1. Compare the performance of two or more carbons under the same dynamic flow conditions.
2. Determine the minimum contact time required to produce the desired quality of carbon column effluent, which is the most important of all design factors.
3. Check the manufacturer's data for head loss at various flow rates through different bed depths.
4. Check the backwash flow rate necessary to expand the carbon bed for cleaning purposes.
5. Establish the carbon dosage required, which will determine the necessary capacity of the carbon regeneration furnaces and auxiliaries.
6. If the overall process or plant flow sheet has not been firmly established, check the effect of various methods of pretreatment (influent water quality) upon carbon column performance, carbon dosage, and overall plant costs.
7. Learn quickly and with a clarity and decisiveness many of the practical advantages and disadvantages that cannot be reached by reading the experiences of others, if a final decision has not been made on the use of upflow or downflow carbon columns or the particle size of carbon to be used.

Figures 7–4 and 7–5 are schematic flow diagrams of upflow and downflow pilot carbon columns. Details of construction of carbon column pilot plants are described in several publications and are more or less apparent in Figures 7–4 and 7–5.

In all of the pilot plant tests the *p*H and temperature should be observed to be certain that they correspond to the values for the full-scale plant operation, since *p*H and temperature have important effects on the carbon treatment.

The first among the things to be learned from pilot tests is a comparison of the different carbons tested. This item overlaps to some extent all of the six which follow, for they all may affect the selection of the carbon. However, the first item is aimed principally at gathering data for the plotting of breakthrough curves. These curves are obtained by passing the water containing the substances to be ad-

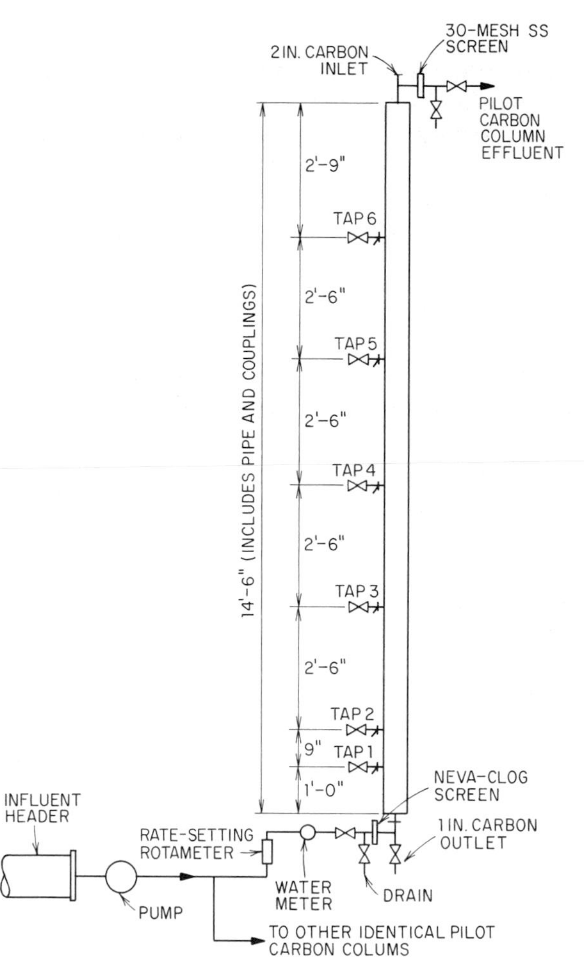

Figure 7–4 Upflow pilot carbon columns. *(Courtesy Clair A. Hill & Assocs.)*

sorbed through a column of carbon. The concentration of the adsorbable substance in the column effluent is plotted as the ordinate against the volume of water treated as the abscissa. This is illustrated for two different carbons, E and F, in Figure 7–6. The breakthrough curve for carbon E is much steeper than for carbon F. This is due to the fact that the rate of adsorption of carbon E is greater than that of carbon F. At the upper right-hand part of the figure, the two curves are terminated at the maximum allowable concentration. Even though the capacity of the two carbons is the same at this point, the carbon with the steepest curve will have the longest service life. This is a

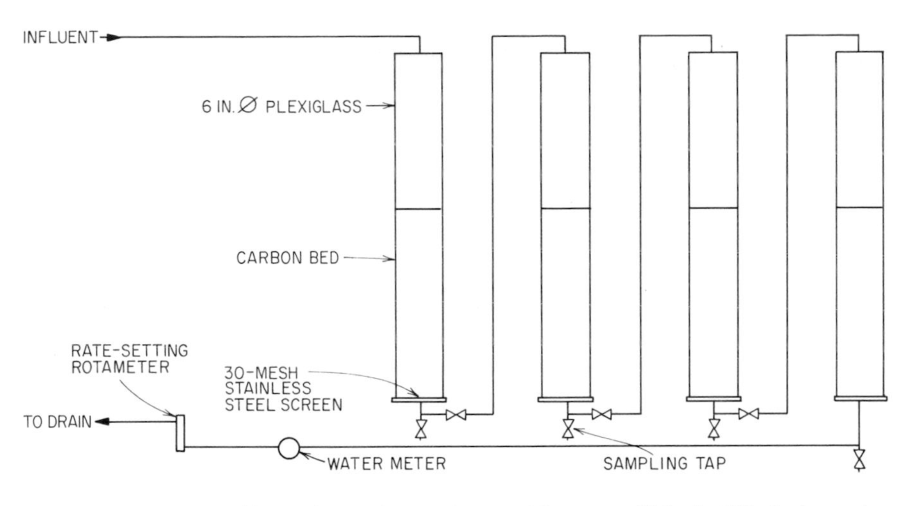

Figure 7–5 Downflow pilot carbon columns. *(Courtesy Clair A. Hill & Assocs.)*

further illustration of the importance of the rate of adsorption by a carbon.

To remove impurities from wastewater, granular activated carbon contact times of 15–35 min have been used. To establish reasonable and proper contact times the objectives of treatment must be kept clearly in mind, and alternative methods of securing the desired water quality must not be overlooked. The desired residual COD concentration is usually the governing factor. For most domestic wastewaters a contact time of 15 minutes gives removals of color, odor, MBAS, and TOC adequate for most reuse applications. Typical COD

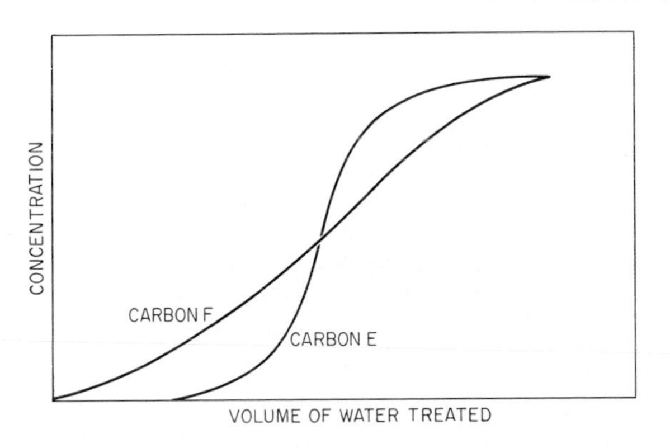

Figure 7–6 Carbon column breakthrough curves.

reductions are about 55 percent. Color and odor are typically reduced to less than current U.S.P.H.S. Drinking Water Standards. If an influent COD of 24 mg/l is assumed, which is possible with proper pretreatment of the wastewater; then the effluent COD from the carbon column is 12 mg/l. This is an acceptable COD value for most purposes, and the necessity for higher removals would require rather strong justification in view of the fact that 72 percent reduction in applied COD may require about 45 min of carbon contact. If a COD content of 24 mg/l again is assumed in the influent, then the effluent concentration at this longer contact time will be about 7 mg/l. Thus to reduce the residual COD from 12–7 mg/l requires tripling the contact time, the volume of the carbon columns, and the total pounds of carbon needed. This difference in quality is not significant under most circumstances. However, if less pretreatment is provided than is assumed in the above example, and if the influent concentration of dissolved COD is, say 60 mg/l, then longer carbon contact times very well may be required, and the additional costs for COD reduction by greater carbon contact must be compared to those for better pretreatment. In most cases it is much cheaper to provide the necessary pretreatment than to install the additional carbon columns and auxiliaries. Lighter loading of the carbon, which is provided by good pretreatment, also reduces the time required for regeneration of the spent carbon and carbon regeneration losses. The optimum contact time is that which results in the minimum total capital and operating cost over the life of the plant. Contact times less than the optimum reduce the capital costs but produce excessive operating costs. Contact times greater than the optimum reduce operating costs but involve excessive capital costs. Unless the plant is to be operated at the same throughput rate, which is equal to the design rate for its entire service life, the optimum contact time provided should be at the *average* rate of operation of the plant during its lifetime rather than at the *maximum* rate. This differs from the sizing of a settling basin, for example, which must be designed for maximum flow rate, because it will not produce the necessary treatment result at higher rates. Carbon columns will continue to function satisfactorily even at reduced contact times, but the operating costs go up. Thus, the optimum carbon column contact time should be based on the average rate of flow over the service life of the carbon facilities rather than the maximum rate of flow expected. This difference from the usual methods for sizing other plant units, if missed, will result in oversized carbon columns. Obviously, the optimum carbon contact time can be provided in several combinations of bed depth and flow rate.

However, carbon efficiency is relatively unaffected by flow rates in the usual range of 4–10 gpm/ft^2, and it is not necessary to use exactly the same depth to cross-section ratio in plant design as in the pilot plant. The configuration of full-scale carbon beds or columns is discussed in detail later in this chapter.

The head loss in inches of water during the carbon service cycle varies from about 2 in./ft of carbon depth (8 × 30 mesh particle size) at 4 gpm/ft^3 to about 5 in. at 10 gpm/ft^2. To expand a bed of 8 × 30 mesh carbon by 10 percent requires a flow of about 10 gpm/ft^2. These typical values (at 72°F) can be field-checked quite easily by use of a carbon column pilot plant.

The pilot test results will also allow a determination of the carbon dosage in terms of pounds of carbon per million gallons of wastewater. Obviously, the dosage will vary with the kind and concentrations of impurities contained in the wastewater. More importantly, the dosage may vary from 200–1600 lb/mg depending upon the degree of pretreatment provided for the wastewater before its application to the carbon. Because granular activated carbon is a relatively expensive material (about 35 cents/lb delivered to Lake Tahoe but less in the East) (1970), it would appear, offhand, that the load of organics applied to the carbon should be minimized by providing a high degree of pretreatment, particularly when there may be loss of adsorptive capacity and mechanical fouling of the carbon by continuous application of water containing suspended or colloidal particles. However, this opinion is not universally accepted, and there is some thought that the carbon should be used to do the maximum amount of purification of which it is capable by applying wastewater with little or no pretreatment. Unfortunately, it is not yet possible to determine with any degree of confidence which approach is best by pilot plant tests. Only full-scale wastewater plant operation under different treatment schemes for a number of years will give a completely satisfactory answer, because some bad effects such as loss of carbon adsorptive capacity may be very slowly accumulative. Caution in this regard lies on the side of using a high degree of pretreatment.

CARBON BED OR COLUMN DESIGN

A great many factors influence the design of contactors for granular activated carbon. One of the most important considerations is to take advantage of what might be called the reserve adsorptive capacity of carbon by using the countercurrent principle. By reserve capacity is meant the ability of a carbon to reduce the color of a water from 100–10 say; when it no longer can adsorb color from a solution of color 10, the same carbon still has a capacity to adsorb color from a solution of color 100 down to 30 say; and so on, until it finally becomes so saturated that it will no longer reduce the 100 color at all, and must be regenerated. By using a countercurrent column, that is, one in which fresh carbon is added at the top and spent carbon is withdrawn from the bottom, and in which the wastewater enters the bottom and exits at the top, all carbon used in the treatment can be fully saturated before it is taken from the column for regeneration. Full countercurrent operation can best be obtained in upflow beds for two reasons: the spent carbon can be easily withdrawn from the bottom of the bed by gravity or by pressurizing the column, and the carbon particles tend to maintain their correct relative position vertically in the bed (even in an expanded bed) because the density of the carbon increases from about 0.48 to about 0.59 as it becomes loaded with organics in its travel down toward the bottom of the bed due to successive withdrawals from the bottom.

Upflow carbon columns for full countercurrent operation may be one of two types, packed-bed or expanded bed. Packed-bed upflow columns are suitable only for low-turbidity waters, that is, waters having a turbidity of 2.5 JU or less. Carbons finer than 8 × 30 mesh should not be used for upflow packed beds because of plugging and high head loss problems. Expanded bed upflow columns may be used with wastewaters containing suspended solids (Hopkins, Weber, and Bloom, 1968). The water must be passed through at a velocity sufficient to expand the bed about 10 percent so that the bed will be self-cleaning. A bed originally expanded hydraulically 10 percent, when restricted by accumulated solids, will further expand so as to continuously flush the solids through the bed and thus avoid increased head loss. For 8 × 30 mesh carbon, this means a flow of about 10 gpm/ft^2 at 72°F. However, 12 × 40 mesh carbon may be used in expanded beds, and 10 percent expansion can be obtained at flows of about 5 gpm/ft^2 at 72°F. Two arguments which have been advanced against the use of upflow packed beds are (1) the plugging of the bottom of the bed and difficulty of cleaning this part of the bed, and (2) the difficulty of installing carbon retaining screens in the top of beds greater than 20 ft in diameter. These arguments are no longer valid. The first problem, that of plugging, can be overcome by using the upflow expanded bed, or by applying low turbidity water. The second can be overcome by using a series of well screens interconnected by a pipe header rather than the old-style continuous fabric flat screens over the entire top cross section of bed.

The only reason for using a downflow bed is to make it possible to use the carbon for two purposes, adsorption of organics and removal by filtration of suspended and flocculated materials. The principal advantage of this dual use of granular carbon is that of some reduction in capital cost. This economic gain is offset to an extent not now fully known by loss of efficiency in both filtration and adsorption, and perhaps also by higher operating costs. O'Farrell, *et al.* (1969) report that a granular carbon adsorption system removed 75 percent of the residual TOC in a coagulated, settled, and filtrated secondary effluent when the applied turbidity was 1–2 JU but only 63 percent from effluents with higher turbidity. The sacrifice in finished water quality which results from combined adsorption-filtration by carbon may or may not be a factor depending upon the quality required under different circumstances. Downflow beds may be fixed either in parallel or in series. Most series designs consist of either two or three beds in series. Valves and piping are provided to permit each bed to be operated in any position in the series sequence, thus giving a pseudo-countercurrent operation. More than three beds in series are seldom used because the cost of valves and piping required becomes excessive.

Often the choice in carbon column configuration lies between single upflow carbon columns, either packed or expanded, operated in parallel preceded by mixed-media filtration, versus a group of two-stage series downflow carbon beds which are dual-purpose filter-contactors. It is always attractive to combine two processes into a single structure, since it offers obvious savings in capital costs. As a general rule, the combined operation is more complex than for either of the two separately, some flexibility and control of operation are lost, and some efficiency of each process must be sacrificed. In the case of combination filter-contractor beds of carbon, all of these things apply. Some savings in capital cost may be realized, but this is the only advantage of this scheme. Equipment for automatic operation of mixed-media filters is standard, highly developed, reliable, and amounts to a satisfactory method of control. Operation and control of upflow countercurrent carbon columns following mixed-media filters or separation beds are best done by simple, foolproof, manual controls. The infrequency of operation does not warrant automatic control. Except for occasional flow reversal, valves serving separate carbon columns are usually operated only during withdrawal and replacement of carbon for regeneration, perhaps once every 4–6 weeks. Valve movement for this purpose is best done manually with operator attendance and observation, and is not amenable to good automatic control. As compared to the simple, reliable, automatic control of the mixed-media beds and manual control of separate upflow countercurrent carbon columns, the logistics, control, and operation of combined carbon filter-contactors can be a complex operational nightmare subject to complete failure and plant shutdown. There are several reasons for this. First of all, the carbon filter-contactor is basically a surface-type, and, as such, is subject to all the well-known shortcomings of surface filters in processing sewage. Such types have never been more than moderately acceptable for filtration of wastewater. Upsets in pretreatment which produce sudden increases in suspended solids or turbidity can completely blind the surface of a single media bed such as a carbon filter-contactor, and it must be backwashed to be restored to service. With two-stage-series carbon filter-contactors the usual arrangement is to have a minimum of six vessels, or three pairs. Two beds are always out of service for regeneration. One bed is being evacuated of spent carbon, and the other is being filled with regenerated carbon. The other two parts of filter-contactors are in service except when one pair is in the backwash cycle. When there is a pretreatment upset and both pairs of beds are surface blinded and out of service for backwashing, then the entire plant is down until at least one pair can be backwashed. If the upset in applied water quality continues for any length of time —say an hour or two, which is not uncommon—then the supply of high-quality backwash water which is necessary for backwashing carbon filter-contactors can well be completely exhausted, the entire plant operation is down indefinitely, and the carbon filter-contactors must be bypassed.

In wastewater plants, it is not necessary to use final effluent or other high-quality water for backwashing mixed-media filters; chemical clarifier effluent is just as satisfactory. Separate filters can be backwashed and placed back on the line in 20 min or less, while a carbon filter-contactor may require 60–90 min, and proportionally more water which must be of high quality to avoid plugging the bottom of the deep bed and saturating it with adsorbed organics. The mixed-media beds would protect the separate upflow countercurrent columns completely from the pretreatment upset, and normal plant operations would continue uninterrupted. With upflow columns, no spare beds are needed, because carbon is withdrawn for regeneration while the column is in service. Unfortunately, this very important practical factor of lack of plant reliability can be completely missed in pilot plant operations, because such plants usually operate at constant flow rates and under other conditions which protect them against pretreatment upsets. Even if upsets do occur they are not serious, since pilot plants do not have to operate continuously, and its vital importance to full plant-scale operations may be discounted or go unrecognized. The common upsets in secondary activated sludge treatment cause no serious problem in conventional plants and often go unnoticed. These same upsets can create impossible operating conditions for surface-type filters which follow, unless some special intermediate protective treatment is inserted into the flow sheet, such as a settling basin with a very low—say 500–600 gpd/ft^2—overflow rate at *peak* hourly plant flows. Considering the greater process reliability, lack of interruption of plant operation, increased carbon efficiency, and decreased carbon regeneration losses provided by mixed-media filtration and upflow countercurrent carbon columns as compared to combination downflow carbon filter-contactors, the choice seems clearly in favor of the former. The savings from use of a single vessel for filtration-carbon contact will probably be more than offset by the necessity to protect these units against pretreatment upsets by construction of an extra or oversized settling basin. In addition, because full-scale, mixed-media filters followed by upflow countercurrent carbon columns have been in continuous uninterrupted operations at South Tahoe from 1968 to 1970, this system is on a firmer, more fully developed basis.

A second consideration in the design of carbon beds is the bed depth to cross-sectional area ratio. With poor distribution and collection arrangements, it is advantageous to use a high ratio, say 2:1, 4:1, or 10:1. However, with properly designed inlet and outlet, and good flow distribution, almost any ratio is satisfactory, even less than 1:1. The problem is almost identical to that of flow equalization through a sand filter. Fortunately, all granular beds act as good distributors and equalizers of flow. High aspect (depth to diameter) ratio columns and columns with small cross-sectional areas are more readily fitted with conical or pyramidal bottoms, which facilitate carbon removal. One factor influencing the overall height of carbon vessels is the climate. In freezing climates the columns are usually housed within a building. A 3-story building (say 35 ft high) will accommodate a carbon column 24 ft tall with an effective carbon depth of about 15 ft, and it also will accommodate the carbon regeneration equipment. In mild climates, carbon columns are not housed and may be as tall as 40–60 ft. Tall, thin columns operate at higher superficial velocities and are deeper. Thus they operate at higher head losses or greater pressure drops than do low-aspect ratio beds. Total plant capacity also influences carbon bed design. For plants of 20 mgd capacity or under, factory-built steel vessels are a good answer, but for large plants larger individual units of poured concrete construction are indicated. The largest completely shop fabricated and tested steel column vessel which can be shipped by rail or truck is about 12 ft in diameter. This means that a single vessel of this kind can handle a flow of about 1 mgd. Obviously, in plants designed for flows greater than about 20 mgd, larger units will effect economies in both construction and operation.

Carbon easily moves down a 60-deg slope, and will move down a 45-deg slope. Small-diameter (8-ft or less) columns are usually fitted with 60-deg tops and bottoms, while larger diameter (9–12-ft) columns may use 45-deg slopes in order to keep the overall height of the column within reasonable limits. Arching or bridging effects are less of a problem in large-diameter columns. When granular carbon is being withdrawn from a column it tends to "rathole," that is, a small-diameter cylinder of carbon directly above the outlet tends to move and to leave in place the material at the outside face of the column. Uniform carbon removal across the entire column cross-section is desired, of course, in order to remove only the bottom, saturated carbon, rather than a mixture of spent carbon and carbon which still has adsorptive capacity for organics at the influent concentration. An upward flow of water in the column at a rate of 8–10 gpm/ft^2 during carbon withdrawal assists greatly in getting the desired uniform removal of carbon from the column through the bottom outlet.

Any vessel which may contain partially dewatered carbon must be protected against corrosion. Dry carbon is not corrosive and neither is carbon in water slurry, but partially dewatered carbon is extremely corrosive and will pit mild steel by electrolytic corrosion to a depth of ¼ in. in a year. Stainless steel is a good but expensive material to use in contact with moist carbon. Also, steel can be adequately protected by use of a good coal-tar epoxy paint. A minimum of three coats to a total thickness of 18–24 mils is recommended. Steel tanks can be equipped with properly designed cathodic protection systems for control of corrosion. A very liberal extra thickness (about 3/16 in.) of steel should be allowed for corrosion above structural or pressure requirements.

The problems of collecting and distributing the flow of water at both top and bottom of carbon columns is similar to those encountered in the design of sand filters. Well screens made of 304 stainless steel are highly recommended for both inlet and outlet service. Referring to Table 7–2, it is seen that a screen slot size 20 (0.020-in. opening) will retain 8 ×30 mesh carbon, and screen slot size 14 (0.014-in. opening) will retain 12 × 40 mesh carbon. It is recommended that the gross area of well screen provided be at least 13 sq ft/mgd of flow.

Table 7-2 Opening Sizes for U.S. Sieves.

U.S. Sieve No.	*Sieve Opening (mm)*	*Sieve Opening (in.)*
3½	5.660	0.223
4	4.760	0.187
5	4.000	0.157
6	3.360	0.132
7	2.830	0.111
8	2.380	0.0937
10	2.000	0.0787
12	1.680	0.0661
14	1.410	0.0555
16	1.190	0.0469
18	1.000	0.0394
20	0.840	0.0331
25	0.710	0.0280
30	0.590	0.0232
35	0.500	0.0197
40	0.420	0.0165
45	0.350	0.0138
50	0.297	0.0117

Figures 7–7 and 7–8 are a cross section and a photograph of carbon columns at the South Tahoe PUD Water Reclamation Plant. These columns will be used to illustrate certain design features. They are upflow full-countercurrent columns and are usually operated as packed types since the turbidity of the applied water is always less than 2.5 JU and usually less than 0.3 JU. However, as indicated by the carbon level in Figure 7–7, the columns have also been operated at times as expanded beds by withdrawing 10 percent of the carbon. The carbon bed is 12 ft in diameter by an effective depth of 14.5 ft. The cross-sectional area is 113 sq ft. Each column contains about 1,810 cu ft or 24 tons of carbon. At a design capacity of 720 gpm, the flow is 6.4 gpm/ft^2, and the contact time is 17 min. The tanks are steel with coal-tar epoxy lining inside. The carbon is 8 × 30 mesh Calgon Filtrasorb 300, and the inlet and outlet screens are Johnson 304 stainless

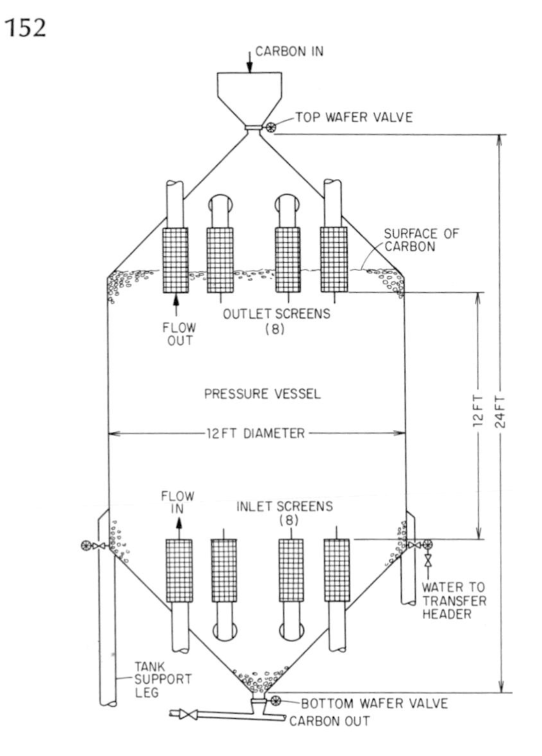

Figure 7–7 Section through carbon column. *(Reprinted from the J. AWWA, 60:1, Jan. 1968, copyright 1968 by AWWA, Inc., 2 Park Ave., New York, 10016)*

Figure 7–8 Carbon columns at South Tahoe. *(Courtesy of Cornell, Howland, Hayes & Merryfield)*

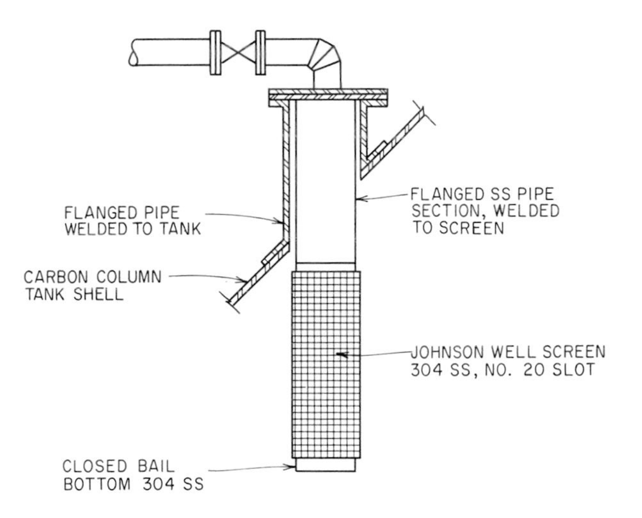

Figure 7–9 Carbon column screen for inlet and outlet. *(Courtesy Cornell, Howland, Hayes & Merryfield)*

steel with No. 20 (0.020-in. openings). The screens are of nominal 12-in. diameter (11.25 in. actual OD) by 2.5 ft long. The total gross screen area is 13.2 sq ft for the eight inlet screens, and is the same for the eight outlet ones. The screens have 75 sq in. of open area per foot of length (Figure 7–9). The top of the column is equipped with a 1-in. combination pressure, pressure-air, vacuum relief valve which has a 1¼-in. Johnson SS well screen, again with 0.020-in. slot openings, inside the tank to prevent carbon plugging of the relief valve (Figure 7–10). Vacuum relief is essential to prevent possible collapse

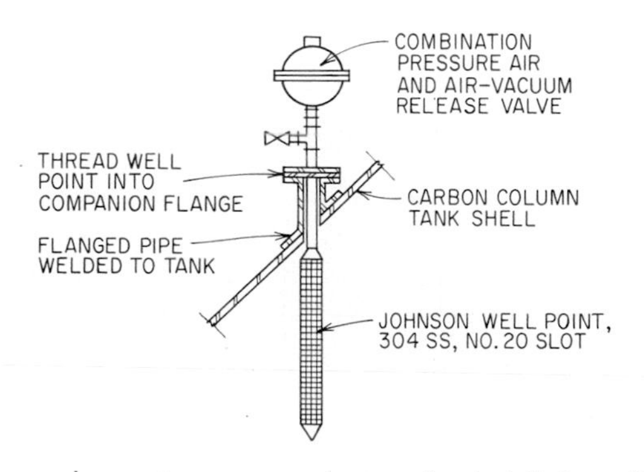

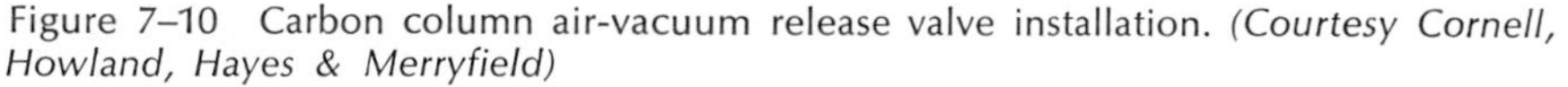

Figure 7–10 Carbon column air-vacuum release valve installation. *(Courtesy Cornell, Howland, Hayes & Merryfield)*

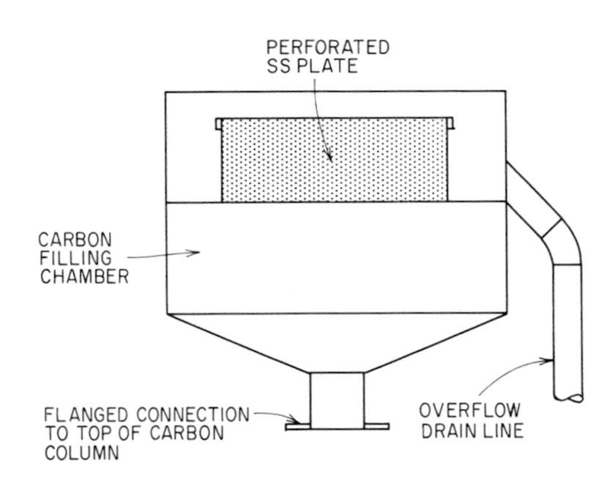

Figure 7–11 Carbon column filling chamber. *(Courtesy Cornell, Howland, Hayes, & Merryfield)*

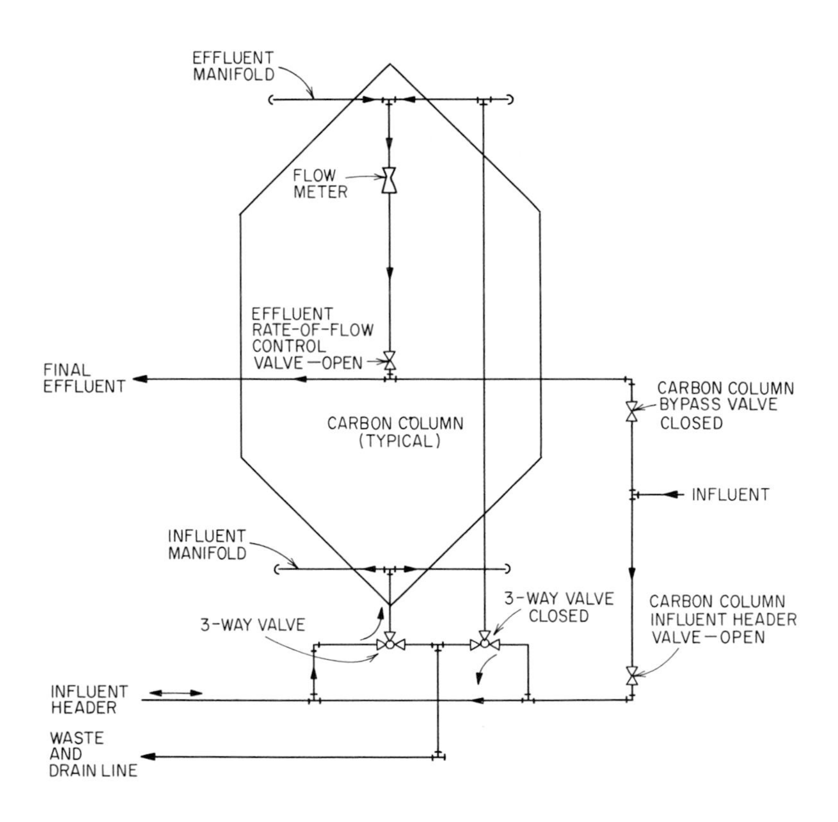

Figure 7–12 Carbon column normal upflow operation. *(Courtesy Clair A. Hill & Assocs.)*

of the steel vessel when it is drained. Pressure air relief prevents entrapment of air when filling the tank. Flow control through the column is provided by a Dall tube and a butterfly valve on the column effluent line. Head loss through each carbon column is continuously monitored and recorded. At the top apex of the column is a column filling chamber (Figure 7–11). The design of this filling chamber allows the column to be completely filled with water when carbon is introduced with the displaced excess water being discharged through the screened overflow drain arrangement as shown. The overflow screen is made from Neva-Clog 24 gage, 304 SS sheet, with perforations 0.045 in. in diameter. As previously stated, the column is operated upflow. However, the direction of flow can be reversed for clearing the top screens and also for flushing out particulate matter from the bottom of the carbon bed. Figure 7–12 is a schematic diagram of the exterior carbon column piping, and Figure 7–13 is a photograph which illustrates carbon column inlet piping.

Figure 7–13 Carbon column inlet header. *(Reprinted from* Water & Wastes Engineering, *6:4, Apr., 1969, p. 36. R. H. Donnelley Corp.)*

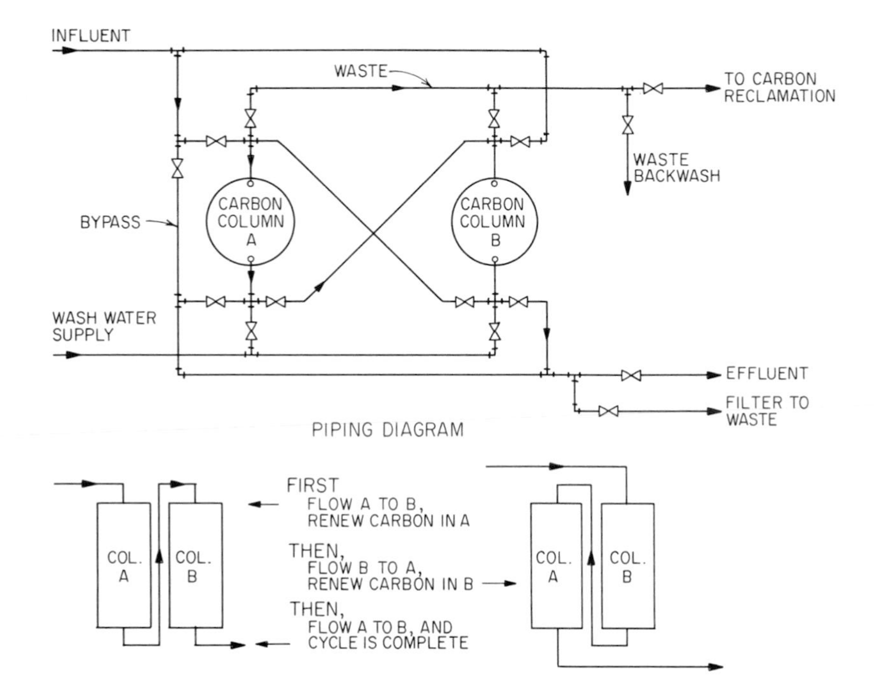

Figure 7–14 Two downflow carbon beds in series. *(Courtesy Cornell, Howland, Hayes & Merryfield)*

Leaving the discussion of upflow countercurrent columns for a moment, downflow carbon beds will be considered. As previously mentioned, the reason for using downflow carbon beds usually is to make use of the carbon both for its adsorptive capacity and its abiilty to act as a filter medium for removal of particulate matter from the wastewater. Downflow carbon beds must be equipped with more elaborate means for backwashing than upflow columns, including installation of surface wash equpment. The underdrain systems used for downflow beds is similar to those used for mixed-media filters and will not be discussed in detail here. Figure 7–14 illustrates a two-bed series downflow system for carbon contact. As indicated, water is first passed down through column A then down through column B. When the carbon in column A is exhausted, the carbon in column B is only partially spent. At this time, all carbon in column A is removed for regeneration, and is replaced with fresh carbon. Column B then becomes the lead column in the series. When the carbon in column B is spent, the carbon is removed for regeneration and is replaced with fresh carbon. This type of operation gives only some of the advantages of countercurrent operation, because only the carbon near the inlet of the lead bed is fully saturated with impurities removed from the water, and some capacity is unused in much of the rest of the carbon sent to regeneration.

CONVEYING GRANULAR CARBON

Granular carbon can be conveyed either by water or by air. However, air transport involves problems of high carbon attrition losses and rapid erosion of pipe and fittings, particularly from sand-blasting effects at pipe bends. Only hydraulic transport is recommended, because it avoids the two difficulties mentioned above and is a cleaner, dust-free operation.

About a gallon of water per pound of carbon is required to form a suitable slurry. As illustrated by Figure 7–15, a dust collector should be installed where bags of dry granular carbon are to be dumped. The minimum diameter for pipelines carrying carbon slurry is 2 in.

Figure 7–15 Use of dust collector when dumping dry carbon. *(Courtesy Cornell, Howland, Hayes & Merryfield)*

Plain steel or cast-iron pipes with extraheavy fittings are satisfactory. Slurry velocities should be not less than 3 ft/sec to prevent settling out of carbon, and not more than 6 ft/sec to minimize carbon abrasion pipe wear. In a 2-in. pipe at the recommended slurry concentration and range of velocities from 25–45 lb/min of carbon can be delivered with pipe friction losses of from 2–10 ft per 100 ft of pipeline.

Both makeup carbon and regenerated carbon must be washed before it is placed in carbon columns in order to remove fine carbon dust and thus avoid plugging and excessive head losses in carbon beds. Figure 7–16 shows a carbon wash tank design. The wash tank is first filled with water, and then the carbon is introduced and washed with the fines passing out through the screens along with the wash water to waste or to plant recycle. Then the wash tank can be pressurized to convey the carbon to the carbon column or other desired point of delivery. The use of a pressurized tank to move carbon slurry is sometimes called the blowcase method.

Carbon slurries can also be transported by use of diaphragm slurry

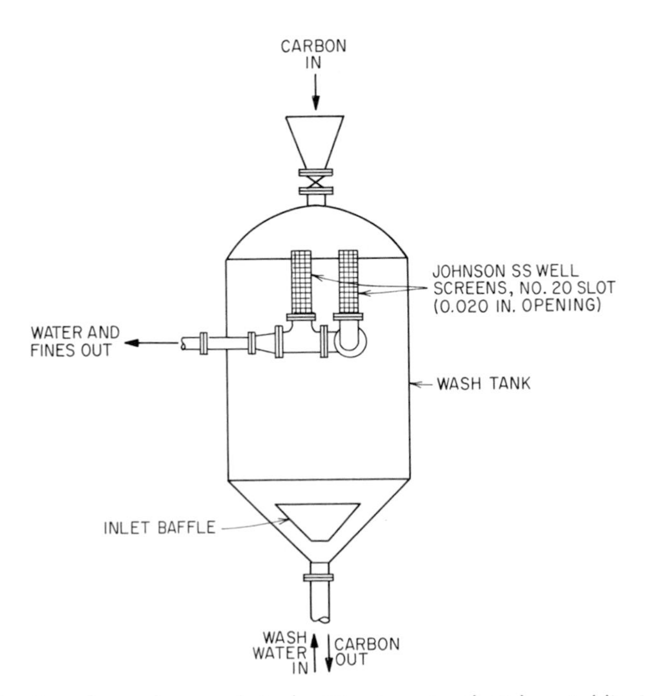

Figure 7–16 Granular carbon wash tank. *(Courtesy South Tahoe Public Utility District)*

pumps (as made by Dorr-Oliver), water jet eductors, or low-speed, rubber-lined, recessed-impeller, centrifugal pumps.

FILLING CARBON COLUMNS

Dry loading of carbon columns is not recommended. Columns should be filled by introducing carbon slurry to a column completely filled with water. Excess water not drawn off by carbon column effluent screens must be drained off through an overflow arrangement as pictured in Figure 7–11, or by opening the column effluent valve in the case of downflow beds. The completely filled bed should then be washed, again to remove carbon fines plus trapped air.

CARBON COLUMN OPERATING RESULTS

In studying operating results of adsorpion of impurities from wastewater by granular activated carbon, it is very worthwhile to review and keep in mind those factors which affect the efficiency of the adsorption process, particularly the effects of pretreatment, and also those factors which may influence plant-scale results as compared to those obtained in pilot plant studies.

The three most important factors which affect final effluent quality and the carbon dosage required are (1) contact time, (2) pretreatment (or applied water quality), and (3) the extent to which countercurrent principles are applied. With proper design of full-scale carbon columns, adsorption should equal that obtained in pilot plant tests. At the South Tahoe Plant, for example, several months of parallel operation between a pilot and a plant column demonstrated that full 100 percent volumetric efficiency was obtained under actual plant conditions, demonstrating that with proper design, full theoretical contact times can be realized. If pilot tests are better than plant results, either poor flow distribution or nonuniform carbon withdrawal from the large units is indicated.

The pretreatment of the wastewater before it is applied to the carbon has a significant effect upon the carbon dosage required and the final effluent quality. Table 7-3 compares dosages and carbon column effluent quality for various pretreatments. Except for the last column of data, which is based on more than 4 years of full-scale plant operations at Tahoe, the figures are estimates based on data obtained from a variety of sources, mostly from pilot plant operations, and they may be subject to considerable change with greater experience in treating stronger wastewaters. However, several things are quite apparent from Table 7–3. With lighter applied loadings, carbon can produce higher quality effluents when they are required. A reduction in applied COD results in lower effluent COD from carbon columns. Carbon can be used as an alternate to other methods of treatment, the choice depending upon the relative economics of the methods to obtain the desired quality. The literature contains many statements to the effect that suspended and colloidal material in water applied to granular carbon have no effect on adsorption. There is some evidence to the contrary, however, noted earlier as reported by O'Farrell, *et al.* (1969). Also at Tahoe, parallel pilot plant tests at 20 minutes contact indicated 20 percent greater adsorption of COD per pound of carbon from chemically coagulated and filtered secondary effluent than from secondary effluent. Further, flushing of accumulated solids from the columns was followed invariably by an abrupt increase in the rate of adsorption. Both of these things strongly suggest loss of capacity due to mechanical fouling of the carbon by solids or colloids in secondary effluent.

Table 7-3 The Effects of Pretreatment on Carbon Dosage and Carbon Column Effluent Quality.

Pretreatment	*Primary*		*Secondary, Plus Plain Filtration*	*Chemically Flocculated and Filtered Secondary Effluent*
Carbon contact	*Downflow series 2 beds 15 min*	*4 beds 30 min*	*Downflow 4 series beds 20 min*	*Upflow countercurrent 17 min*
Carbon dosage lb/mg	1,200	800	500	250
SS, mg/l	10	5	<1	<1
BOD, mg/l	20	10	<1	<1
COD, mg/l	65	45	12	12
TOC, mg/l	20	10	3	3
Color, units	—	—	4	4
Turbidity, JU	—	—	1.5	0.5

Upsets in secondary treatment may cause frequent backwashing of downflow beds receiving secondary effluent directly. Backwashing of downflow beds causes turnover of the beds due to the fact that the high-density (most saturated) particles which are at the inlet (top) of the bed move down when the bed is fluidized, thus giving a reverse countercurrent effect. An upflow bed is resistant to overturn during expansion due to the fact that the high-density saturated particles stay near the inlet, which in this case is the bottom of the bed where they belong for countercurrent operation.

Carbon columns handle shock dissolved organic or hydraulic loads very well, better than most other unit processes for treating wastewater, which is often noted for its wide variations in strength and quantity.

The longest plant experience in treating wastewater in columns of granular activated carbon has been accumulated at South Tahoe. In 1965 operations started with two columns having a total capacity of 2.5 mgd. In 1968 the capacity was increased to 7.5 mgd. Based on successful experience with the original carbon plant, no design changes were made in the second phase of construction. The pretreatment and design of the carbon adsorption system have already been described. Typical analyses of wastewater before and after carbon treatment following 4 years of operation are given in Table 7–4.

Table 7-4 Typical Water Quality Before and After Granular Activated Carbon Treatment at South Tahoe.

	CARBON COLUMN	
Quality Parameter	*Influent*	*Effluent*
BOD (mg/l)	3	<1
COD (mg/l)	24	12
TOC (mg/l)	12	3
MBAS (mg/l)	0.85	0.13
Color (units)	15	4

During the more than 4½ years of carbon treatment, about 3¼ billion gal of wastewater have been passed through the carbon beds. The carbon has removed approximately 90 tons of detergent, 3,100 tons of oxygen-consuming substances, and 260 tons of color bodies, or a total of about 3,450 tons of organic substances. Carbon adsorption has proven to be a very simple and reliable process, and has taken its place in the arsenal of weapons to be used in the fight to control water pollution.

COSTS FOR CARBON ADSORPTION

At design capacity of 7.5 mgd, the capital costs for carbon adsorption at Tahoe on the basis of national average replacement construction costs for 1969 were $623,000, or $84,200/mgd of capacity. The capital costs, based on 25-year amortization at 5-percent interest, amount to $16.30/mg. Carbon operating costs are related to carbon regeneration costs and for that reason are discussed following the next chapter, which is on the subject of carbon regeneration.

References

1. Calgon Corp., *Water and Waste Treatment With Granular Activated Carbon,* General Catalog (1970).
2. Bishop, D. F., *et al.,* "Studies on Activated Carbon Treatment," *Journal Water Pollution Control Federation* (Feb., 1967).
3. Cooper, J. C., and Hager, D. G., "Water Reclamation With Activated Carbon," *Chemical Engineering Progress,* 62:85 (Oct., 1966).
4. Cover, A. E., and Wood, C. D., "Appraisal of Granular Carbon Contactors, Phase III," FWPCA Report No. TWRC–12 (May, 1969).
5. Cover, A. E., and Pieroni, L. J., "Appraisal of Granular Carbon Contactors, Phases I & II," FWPCA Report No. TWRC–11 (May, 1969).
6. Culp, R. L., "Wastewater Reclamation at South Tahoe PUD," *Journal American Water Works Assoc.,* 1968, p. 84.
7. Fornwalt, H. J., and Hutchins, R. A., "Purifying Liquids With Activated Carbon," *Chemical Engineering,* May 9, 1966, p. 155.
8. Hassler, J. W., *Activated Carbon,* Chemical Publishing Co., New York, 1963.
9. Hopkins, C. B., Weber, W. J., and Bloom, R. "A Comparison Of Expanded Bed And Packed Bed Adsorption Systems," FWPCA Report No. TWRC–2, (Dec., 1968).
10. Hager, D. G., and Flentje, M. E., "Removal Of Organic Contaminants by Granular Carbon Filtration," *Journal American Water Works Assoc.,* 1965, p. 1440.
11. Joyce, R. S., Allen, J. B., and Sukenik, V. A., "Treatment of Municipal Wastewater by Packed Activated Carbon Beds," *Journal Water Pollution Control Federation,* 1966, p. 813.
12. Morris, J. C., and Weber, W. J., "Adsorption of Biochemically Resistant Materials From Solution," USPHS AWTR Publication No. 9 (1964).
13. O'Farrell, T. P., Bishop, D. F., and Bennett, S. M., "Advanced Waste Treatment at Washington, D.C." presented at the Sixty-fifth Annual AICHE meeting, Cleveland, Ohio (May, 1969).
14. Parkhurst, J. D., Dryden, F. D., McDermott, G. N., and English, J., "Pomona Activated Carbon Pilot Plant," *Journal Water Pollution Control Federation,* 1967, p. R70.
15. Rizzo, J. L., and Schade, R. E., "Secondary Treatment With Granular Activated Carbon," *Water and Sewage Works,* Aug. 1969, p. 307.
16. Slechta, A. F., and Culp, G. L., "Water Reclamation Studies at The South Tahoe Public Utility District," *Journal Water Pollution Control Federation,* p. 787 (1967).
17. Smith, C. E., and Chapman, R. L., "Recovery of Coagulant, Nitrogen Removal, and Carbon Regeneration in Wastewater Reclamation," Final Report to FWPCA, Demo. Grant WPD–85 (June, 1967).
18. Westvaco Bulletin, *Activated Carbon and Wastewater,* West Virginia Pulp And Paper Co., 1970.

8

Granular Carbon Regeneration

THE REGENERATION PROCESS

In very small plants it may be feasible to use granular carbon on a once-through, throwaway basis, although economics probably would favor the use of powdered carbon in this situation. Small plants can also consider the use of central regeneration facilities if they are available within a reasonable hauling distance. With these minor exceptions, the use of granular activated carbon involves the regeneration and reuse of the carbon. Indeed, the amenability of carbon to regeneration and reuse, and the fact that there are no liquid or solid byproducts from its use requiring disposal are among activated carbon's principal process advantages for removing organics from wastewater.

There are four general methods for reactivating granular carbon: solvent wash, acid or caustic wash, steam reactivation, and thermal regeneration. With use of a solvent which will dissolve the adsorbed material, the absorbate is passed through the carbon bed in the direction opposite that of the service cycle until it is removed. The bed is then drained or purged of the solvent, and the regenerated carbon is ready to go back on-stream. If an acid or caustic is more effective than a solvent in dissolving a particular adsorbate, it may be used in a manner identical to that described for solvent washing. Adsorbates with low boiling temperatures can sometimes be removed by steam. Approximately 3–5 lb of steam per pound of adsorbate is passed through the carbon opposite to normal flow, and is either vented to the atmosphere or condensed and recovered. These first three methods do not appear to have application as the primary methods for regeneration of activated carbon which has been used for the treatment of wastewater. The fourth method, that of thermal regeneration, is universally used for this purpose at the present time.

Juhola and Tupper (1969) have shown that thermal regeneration of granular carbon consists of three basic steps: (1) drying; (2) baking (or pyrolysis) of adsorbates; and (3) activating, by oxidation of the carbon residues from decomposed adsorbates. Drying may be accomplished at 212°F, baking between 212 and 1500°F, and activating at *carbon* temperatures above 1500°F. All of these steps can be carried out in a direct-fired, multiple hearth furnace. Presently this is the best commercially available equipment for regeneration of carbon used in wastewater treatment. The principal disadvantages of present furnace designs, as pointed out by Juhola and Tupper (1969) are the high gas input rate, short activating time, and low gas utilization (about 18 percent). The same investigators also mention a reduction in Iodine Number of about 25–100 units (of a total of about 950) brought about by water quenching of the red hot carbon. Juhola and Tupper recommend as a result of their work, which included investigation of an indirect heated rotary tube furnace, that studies be made "to establish operating limits for engineering of the process and equipment and to subsequently determine the relative economics of regeneration by (1) indirect heated rotary tube furnace, (2) direct fired multiple hearth furnace, (3) a combination of the two, and (4) a two pass regeneration using the multiple hearth furnace." These recommendations certainly should be carried out, for ultimately, persuance of these studies and ideas may lead to development of better equipment and processes for regenerating carbon. However, there is no need to delay projects currently under consideration awaiting perfection of these possible improvements, because multiple hearth furnaces now available are a simple, easily controlled, reliable, practical, and economical means of regenerating carbon.

REGENERATION EQUIPMENT

At South Tahoe, a commercial size B-S-P Corporation multiple hearth furnace has been in operation since 1965 (over 4½ years at this writing in 1970) in the successful regeneration of spent granular carbon used in full-scale wastewater reclamation. This installation is typical for the processing of wastewater, and will be used as an example to illustrate some of the principles involved in design and operation of carbon regeneration facilities. Nichols Engineering Company also makes similar multiple hearth furnaces.

The design capacity of the Tahoe plant is 7.5 mgd. At the present time, the carbon dosage averages about 250 lb/mg. The theoretical required furnace capacity is then, $7.5 \times 250 = 1{,}875$ lb/day. This computation does not allow for furnace downtime or other contingencies. Also, the smallest commercial furnace available is one with a 54-in. diameter, which is rated at 6,000 lb/day. at 7.5 mgd, this furnace would handle carbon dosages up to 800 lb/mg, which gives a comfortable margin of safety. The furnace has satisfactorily regenerated carbon at a feed rate of 4,800 lb of dry carbon per day at temperatures on the fired hearths of 1650–1700°F.

REGENERATION PROCEDURES AT TAHOE

When the carbon column effluent reaches the minimum effluent standards, spent carbon is removed from the columns. These standards are as follows: COD, 15 mg/l (or less than 50 percent of applied COD); color, 5 units; and MBAS, 0.5 mg/l. Before the advent of bio-degradable detergents, ABS was the governing criteria, but now the soft detergents are broken down in pretreatment to the extent that COD controls. Experience in operating the Tahoe plant has shown that the COD breakthrough occurs when the carbon dosage is about 215 lb/mg, but that the carbon is much easier and cheaper to regenerate if withdrawn earlier, at the time the dosage is about 250 lb/mg—so this has become the basic signal for beginning carbon withdrawal. More highly saturated carbon takes two times as long to regenerate.

The carbon columns are pressurized to withdraw the spent carbon. Carbon is withdrawn while the column is in service. Upward flow is increased to about 10 gpm/ft^2 in order to expand the carbon bed slightly and to assist in uniform withdrawal of the spent carbon, that is, to prevent "ratholing" of the carbon. About 11 percent (5200 lb) of the contents of one column are removed at one time and replaced with regenerated carbon.

A flow sheet of the regeneration system is shown in Figure 8-1. Spent carbon taken in slurry form from the columns passes to either of two drain and feed tanks. These tanks are located at an elevation

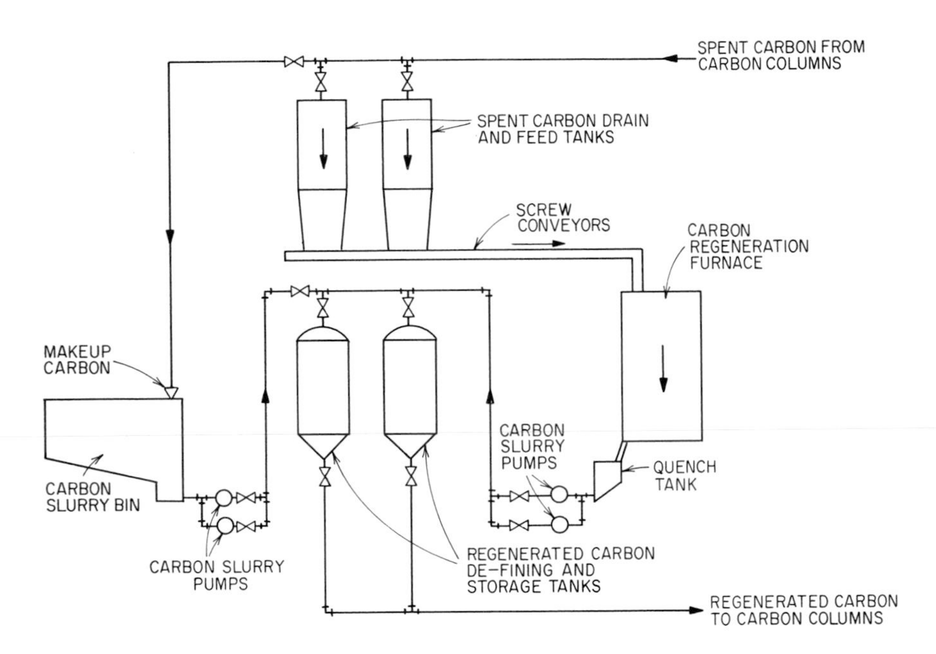

Figure 8–1 Carbon regeneration system. *(Courtesy Clair A. Hill & Assocs.)*

above the top of the furnace. Each dewatering bin will hold a slug of about 200 cu ft (5200 lb dry weight) of carbon, or a total of 400 cu ft (10,400 lb dry weight) in the two tanks. When carbon slurry is deposited in the spent carbon bin the water must be drained off through screens in the tank sides at the bottom to about 50-percent moisture before it can be fed to the carbon regeneration furnace. This takes about 15 min.

The partially dewatered carbon is fed from the drain bins to the furnace by means of a screw conveyor. This conveyor is constructed entirely of 304 SS to resist corrosion. The screw conveyor is equipped with a variable speed drive so that the rate of carbon feed to the furnace can be controlled accurately. The furnace has a turndown ratio of 6 to 1, so that it can operate at rates between 1,000 and 6,000 lb of dry carbon per day. This corresponds to 33.3–200 cu ft/day, or 0.023–0.138 cfm. The 9-in. diameter screw has a 3-in. pitch at the lower (drain bin) end, and a 9-in. pitch at the upper (furnace) end (Figure 8-2). The low-pitch, metering section of the screw under the bin runs full of carbon for accurate feed control, and the higher pitch, conveying section runs only partially full to reduce the total driving torque required. At 100-percent efficiency (no slippage) the

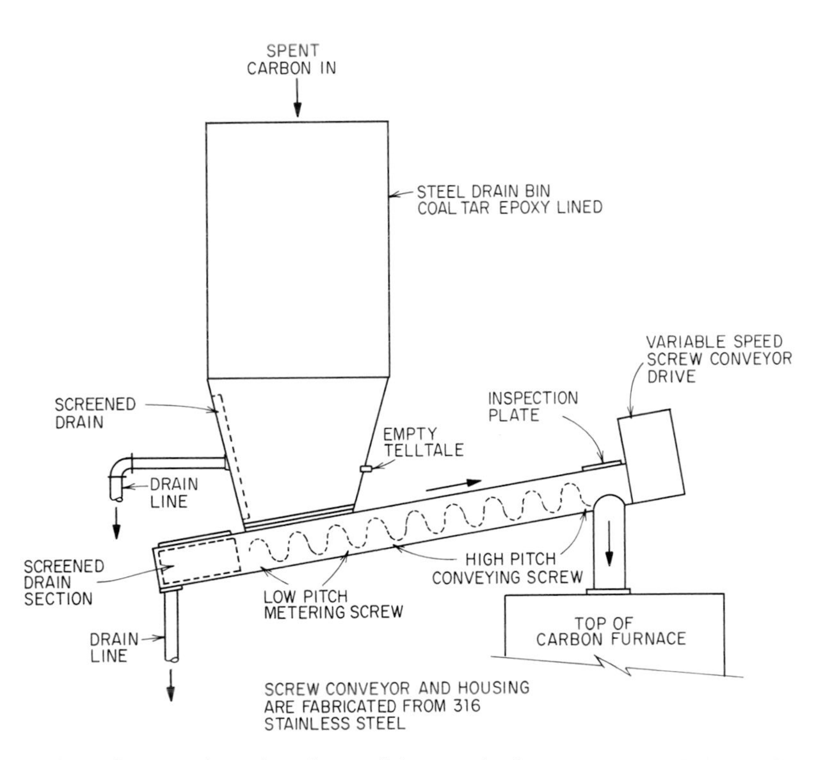

Figure 8–2 Spent carbon drain bin and furnace feed arrangement. *(Courtesy Cornell, Howland, Hayes & Merryfield)*

9-in. pitch section would deliver 0.25 cu ft/rev, and the 3-in. pitch section 0.083 cu ft/rev. At 100-percent efficiency, the screw speed for the 3-in. pitch section should be 0.3–1.8 rpm in order to get the desired rate of carbon delivery to the furnace. The screw is driven through a Winsmith Reducer (952.6–1 reduction) by a Reeves Drive which can be varied from 216–2160 rpm. After reduction, the speed range for the drive is 0.2–2.0 rpm. To deliver 6,000 lb/day then, the 9-in. portion of the screw must have an efficiency of at least 30 percent, and the 3-in.-pitch section an efficiency of 90 percent (10-percent slippage).

Near the bottom of the drain and feed bin is a low-level alarm (Bindicator) which indicates when the spent carbon level reaches the minimum level adequate to maintain a carbon seal in the conveyor and to prevent gases produced in the furnace from escaping into the building.

The spent carbon is discharged into the top of the carbon regen-

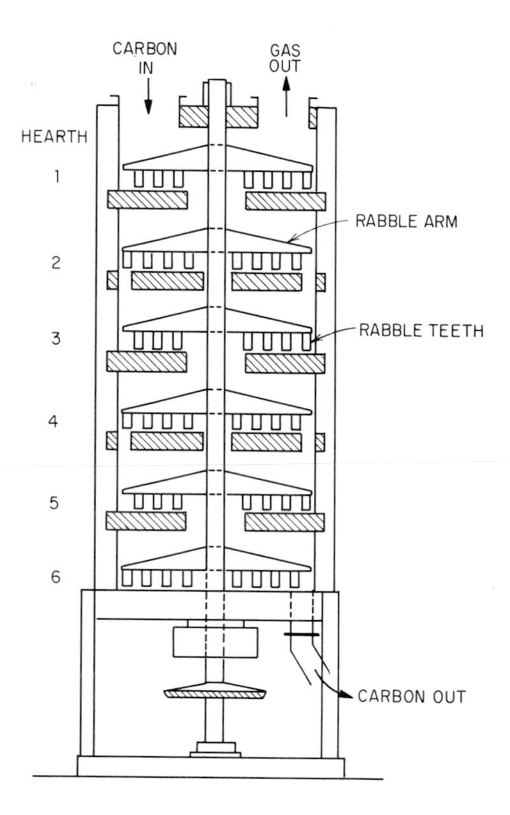

Figure 8–3 Cross-sectional view of multiple hearth furnace.

eration furnace through a 10-in. inlet port. Figure 8-3 is a cross-sectional view of the B-S-P Corporation multiple hearth furnace. This is a 54-in.-diameter, 6-hearth, gas-fired furnace with propane standby. The exhaust gases pass through an afterburner and a wet scrubber to eliminate air pollution. The carbon is moved downward through the six hearths (numbered 1 through 6 from top to bottom) by stainless steel rabble arms.

There are two burners on both hearths 4 and 6. The temperatures on these two hearths are independently controlled by an automatic temperature controller which maintains the temperature within 10°F of the desired temperature. A proportional flow meter is provided in each of the air-gas mixture lines to the burners to insure a constant percentage of excess oxygen. In addition, steam can be added to hearths 4 and 6.

An ultraviolet scanner is provided on each of the burners. In case of a flame-out, the furnace is automatically shut down. Other safety features which shut down the furnace are high or low gas pressure; high combustion air pressure; low scrubber water pressure; high stack gas temperature; and draft fan or shaft cooling air fan not operating. In addition, the shaft cooling air fan is on a standby power circuit to avoid possible damage to the furnace rabble arms in the event of an electrical outage.

Temperatures on the various hearths are about as follows: No. 1, 800°; No. 2, 1000°; No. 3, 1300°; No. 4, 1680°; No. 5, 1600°; and No. 6, 1680°F. In the furnace the carbon is thermally regenerated in the presence of steam. The heat vaporizes and drives off as gas the impurities which are adsorbed on the carbon, and restores the carbon essentially to pristine activity. The addition of steam on hearths 4 and 6 gives a more uniform distribution of temperatures throughout the furnace. It also reduces the apparent density and increases the Iodine Number of the regenerated carbon. About 1 lb of steam per pound of dry carbon is used.

The gases produced from the carbon regeneration leave the top of the furnace and enter an afterburner. When needed, the afterburner is operated at 1200°F to burn volatile and noxious gases. Under most conditions the afterburner is not needed, and wet scrubbing of the stack gas is sufficient. The wet scrubber removes carbon dust and odorous substances, and the stack gas poses no air pollution problems.

The regenerated carbon is discharged from the bottom of the furnace into a small quench tank. Figures 8-4 and 8-5 are a drawing and a photograph, respectively, of a quench tank arrangement. Note that the quench tank has two separate pump suction lines from it, each with several water jets strategically placed to keep the carbon moving in the quench tank and connecting pipelines. Considerable agitation of the carbon slurry is necessary at this point to prevent buildup of carbon in the quench tank and plugging of the pump suction lines. From the quench tank the carbon slurry is pumped to the regenerated carbon storage tanks or wash tanks. Two diaphragm slurry pumps manufactured by Dorr-Oliver, Inc., each having a capacity of 3–20 gpm are used to pump the carbon. The pumps discharge through a 1-in. line to the storage or wash tanks. The process of washing out the fines from the regenerated carbon is the same as that described previously for new carbon when filling carbon columns for the first time. Following the washing, the wash tank is pressurized and the regenerated carbon is forced in water slurry through a 2-in. line to the top of the carbon columns. Makeup carbon is added to the system to replace the carbon lost in the regeneration process by

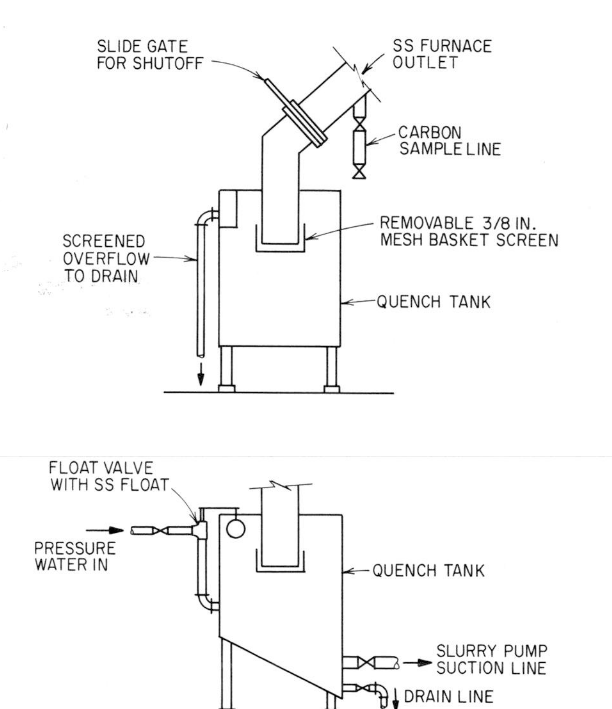

Figure 8–4 Carbon quench tank. *(Courtesy Howland, Hayes & Merryfield)*

dumping bags of new carbon into a concrete slurry bin. The makeup carbon is pumped from this bin by two diaphragm slurry pumps to the wash tanks where it is washed prior to being placed in the carbon columns. The carbon bag dump is equipped with a hinged cover and dust collector as already pictured.

Operation of the furnace is controlled routinely by the apparent density of the carbon, supplemented at times by tests of Iodine Number, adsorption isotherms, or other laboratory tests to check the activity of the regenerated carbon. The carbon is regenerated to an apparent density of 0.48. If the apparent density is greater than 0.49, the furnace feed rate is too high or the temperature too low. If the apparent density is less than 0.48, the furnace temperature is too high or the feed rate too low and chances are that some carbon is being burned. Progressive decreases in the Iodine Numbers of regenerated carbon indicate that the regenerating temperature is too low and that adsorbates are being carbonized and left in the original pore structure. The temperature of the carbon itself during regenera-

Figure 8–5 Carbon furnace discharge into quench tank with connection to slurry pumps in pit below floor. *(Courtesy Cornell, Howland, Hayes & Merryfield)*

tion must be at least 1500°F to avoid this deterioration, and the indicated correction is to raise furnace temperatures. Temperatures as high as 1750 or 1800°F may be required, and full restoration of the Iodine Number may not be obtained until after several successive regenerations. This might be avoided by using the special equipment required to measure the actual carbon temperatures during regeneration to be sure that they are not less than 1500°F. The procedures for performing the regeneration control tests such as AD (apparent density), Iodine Number, and adsorption isotherms, are given in Chapter 11.

Figure 8-6 is a photograph of the complete carbon regeneration system at South Tahoe.

Figure 8–6 Carbon regeneration system. *(Reprinted from* Journal of the American Water Works Assoc., *60:1 (Jan., 1968). Copyright 1968 by AWWA Inc., 2 Park Ave., New York, 10016)*

RESULTS OF REGENERATION

The accumulation of full scale plant data on regenerated carbon is an extremely slow process because of the long period of time between regeneration cycles. Eight to ten slugs of carbon must be withdrawn from each carbon column per cycle, and the period between slugs is from four to six weeks depending upon plant flows. After 4½ years of operation at Tahoe, the oldest carbon in the plant is in only the fourth cycle of reuse. To date the carbon has been regenerated to original apparent density and nearly virginal Iodine Number. No appreciable ash buildup has been detected. In accelerated laboratory tests, granular carbon saturated with organics from wastewater has been regenerated through 20 cycles of use.

Table 8-1 summarizes the results of plant-scale regeneration at Tahoe through the third cycle of reuse. Table 8-2 presents more detailed information for each cycle of regeneration, including the regeneration of fourth-cycle carbon.

Table 8-1 South Tahoe Water Reclamation Plant—Plant Scale Regeneration Through Three Cycles of Reuse.

Carbon Property	*Virginal Carbon*	*Spent Carbon*	*Carbon After 1st Regeneration (1965)*	*Carbon After 2nd Regeneration (1967)*	*Carbon After 3rd Regeneration (1969)*
Apparent density	0.48	0.52–0.59	0.47–0.50	0.47–0.50	0.48–0.49
Iodine number	935	550–843	900–969	782–933	743–875
Ash (%)	5.0	4.5	4.7	4.8–5.9	6.6–8.2

Typically the furnace was operated at 1650°F on the fourth and sixth hearths. The burners on hearth 6 were set to deliver 2 percent excess oxygen and the burners on hearth 4 were set to deliver 4 percent. The carbon feed rate is varied, of course, to suit spent carbon properties, but averages about 240 lb/hr. The moisture in spent carbon fed to the furnace varies from 40–44 percent. The total steam added to the furnace averages about 1 lb per pound of dry carbon. When steam is added, temperatures on hearths 1, 2, and 3 rise about 200°F, thus giving a more uniform distribution of temperatures through the furnace. At the same temperature and feed rate, steam addition markedly reduces apparent density (0.50–0.49) and increases the Iodine Number (880–910). The Btu input to the furnace varies from 2,500–5,000/lb of carbon.

Carbon losses and labor are the two major items of cost in carbon regeneration. The cost of makeup carbon may be as much as 50 percent of the total cost of carbon treatment. Minimizing carbon losses is essential to economic carbon regeneration. Measuring short-term carbon losses precisely is difficult. The most accurate method of determination is the measurement over a long period of time of the quantities of makeup carbon required on a weight basis to maintain the proper level in the carbon columns. Pinpointing the location where carbon losses occur is not easy. Juhola and Tupper (1970) determined the mechanical attrition losses in an externally heated ro-

Table 8-2 Carbon Regeneration Data, South Tahoe Public Utility District.

Date	Regeneration Cycle	Carbon	Carbon Feed (dry lb/hr)	Btu (per lb)	Apparent Density	Iodine No.	Ash (%)	FURNACE HEARTH TEMPERATURES, °F 1	2	3	4	5	6
11-18-69	First	Spent	217	3,200	0.520	674	5.5	635	805	1115	1620	1500	1640
		Regen.			0.479	818	6.3						
11-20-69	First	Spent	158	4,450	0.531	658	6.4	595	780	1120	1630	1505	1640
		Regen.			0.485	788	6.7						
1-12-70	First	Spent	202	3,315	0.560	626	6.2	720	860	1110	1645	1500	1670
		Regen.			0.482	894	6.7						
2-16-70	Second	Spent	176	3,700	0.577	539	6.4	765	935	1200	1695	1550	1700
		Regen.			0.486	860	7.2						
2-20-70	Second	Spent	185	2,890	0.571	546	6.1	570	835	1120	1685	1550	1685
		Regen.			0.483	830	6.7						
2-21-70	Second	Spent	162	3,530	0.570	607	5.7	650	865	1230	1675	1540	1675
		Regen.			0.479	874	6.3						
1- 3-70	Third	Spent	99	5,070	0.553	599	7.8	730	970	1310	1635	1490	1635
		Regen.			0.501	789	7.3						
1- 7-70	Third	Spent	177	4,150	0.515	676	7.6	805	1000	1320	1630	1460	1650
		Regen.			0.488	823	7.3						
2-12-70	Third	Spent	119	5,230	0.580	524	6.5	835	1045	1330	1695	1525	1690
		Regen.			0.488	816	7.7						
2-14-70	Fourth	Spent	116	5,070	0.580	528	6.4	855	1100	1420	1695	1540	1695
		Regen.			0.487	820	7.1						

tating tube regenerator to be about 0.2 percentfi by operating the system without heat. Observations at Tahoe also indicate that mechanical attrition losses are low. In the Tahoe plant carbon losses have varied from 2–12 percent/cycle. The average loss is about 5 percent/-cycle. The carbon loss varies with the feed rate to the furnace, losses increasing at increased feed rates and highest temperatures. In some sugar refineries carbon is precooled before quenching in an attempt to reduce carbon losses. It is claimed to be effective but no definitive data is available. Burning of carbon in the furnace during operation, and particularly during start-up and shutdown is a major factor. Continuous furnace operation will avoid the start-up and shutdown losses, of course, and these losses can be minimized by operating as long as possible during a single regeneration "campaign."

The expanded 7½ mgd plant at Tahoe completed its first full year of carbon operation in March, 1969. During the previous year of operation, the plant inflow and output was 673 million gal. A total of 85 tons of carbon were regenerated, and the average carbon dose was 228 lb/mg. The actual volume of water passed through the carbon was 1,104 mg. This is greater than the plant flow due to recycling of various waste process streams within the plant. Table 8-3 shows the effect of carbon treatment on the wastewater, based on analyses of 24-hour composite samples and duplicate COD tests.

Table 8-3 Results of Carbon Treatment at Tahoe March 1, 1968 to February 28, 1969.

	Influent (mg/1)	*Effluent (mg/1)*	*Applied (lb)*	*Removed (lb)*	*Pounds Removed per Pound Applied*
COD	20.7	11.4	190,500	85,500	0.45
MBAS	0.46	0.13	4,010	2,790	0.70

Costs

The actual costs for carbon regeneration for the same 12-month period are given in Table 8-4.

The cost of regenerated carbon of $0.0323/lb is less than 10 percent of the cost of new carbon, which is about $0.35/lb.

Using this regeneration cost, then, the total cost of carbon treatment at the plant design capacity of 7.5 mgd is given by Table 8-5.

Table 8-4 Actual Tahoe Costs for Carbon Regeneration March 1, 1968 to February 28, 1969.

Item	*Total Cost ($)*	*Unit Cost ($/lb)*
Labor	4,090	0.0241
Fuel	355	0.0021
Power	141	0.0008
Maintenance	905	0.0053
Total	5,491	0.0323

Table 8-5 Total Carbon Treatment Costs at Tahoe at 7.5-mgd Rate Based on Actual Costs for March 1, 1968, to February 28, 1969.

Item	*Cost per mg ($)*
Capital cost	21.10
Regeneration	7.36
Makeup carbon	6.76
Maintenance	0.89
Total cost per million gallons	36.11

The capital cost used represents the actual construction costs at South Tahoe of $849,772, amortized at 4.4 percent for 25 years. The other items in Table 8-4 were computed from actual plant expense records, as follows: Labor $5.16 per hour, fuel $0.054 per therm, power $0.0085 per kwh plus $2.00 per kw demand charge, and makeup carbon at an average loss of 7.6 percent per cycle. It should be noted that the carbon loss in the given year was considerably above the average of 5 percent for the entire 4½ years of operation to date.

References

1. Culp, G. L., and Slechta, A. F., "Plant Scale Reactivation and Reuse of Carbon in Wastewater Reclamation," *Water and Sewage Works,* Nov., 1966, p. 425.
2. Culp, R. L. "Wastewater Reclamation at South Tahoe PUD," JAWWA, p. 84, 1968.
3. Hassler, J. W., *Activated Carbon,* Chemical Publishing Co., New York, 1963.
4. Juhola, A. J., and Tupper, F., "Laboratory Investigation of the Regeneration of Spent Granular Activated Carbon," FWPCA Report No. TWRC–7 (Feb., 1969).
5. Slechta, A. F., and Culp, G. L. "Water Reclamation Studies at the South Tahoe PUD," *Journal Water Pollution Control Federation,* 1967, p. 787.

6. Smith, C. E., and Chapman, R. L., "Recovery of Coagulant, Nitrogen Removal, and Carbon Regeneration in Wastewater Reclamation," Final Report to FWPCA, Demo. Grant WPD-85 (June, 1967).
7. Suhr, L. G., Chapman, R. L., and Culp, R. L., "Operations Manual For 7.5 MGD Water Reclamation Plant, South Tahoe PUD" (Sept., 1967).

9
Solids Handling

INTEGRATION OF PRIMARY AND SECONDARY SEWAGE SLUDGE PROCESSING INTO DESIGN OF ADVANCED WASTEWATER TREATMENT PLANTS

In conventional primary and secondary sewage treatment plants, processing and disposal of primary and secondary sludge, particularly waste activated sludge, poses some difficult problems which in the past quite often have not been satisfactorily resolved in plant design. In advanced wastewater treatment, the use of chemicals produces great quantities of a different type of sludge to be handled, further complicating the sludge problem.

Because sewage sludges are often treated with chemicals to improve their dewatering and handling characteristics, it would seem that it might be advantageous to settle the sewage and chemical sludges together in a single settling basin or to mix the two sludges. Unfortunately, exactly the opposite is sometimes true. That is, it is often better to keep the sludges apart to avoid mixing the two. Mixtures of chemical and sewage sludges are often much more difficult to dewater than is either chemical sludge or sewage sludge when each is handled separately.

The experience to date in the operation of full-scale plants has shown indisputably that the key to successful and economical advanced wastewater treatment plants lies in the methods used for sludge handling. No other factor is so important in the design of advanced wastewater treatment facilities. There are many alternate unit processes and numerous combinations thereof which will produce the required quality of treated water at approximately equal costs. There are comparatively few methods for satisfactory and economic processing of sludge. Therefore, unit liquid treatment processes and plant flow sheets necessarily must be selected on the basis of their effects on the ease and cost of sludge handling.

In considering the sludge problem, the first question to be answered is whether or not there is a need to dewater the sludges produced. The answer to this question depends on local conditions affecting sludge disposal. If the situation is such that the inexpensive, but crude, expedient of land disposal of wet sludge is permissible, then there is no need to dewater it. However, in most locations where advanced wastewater treatment is needed, it is also very likely that better methods of sludge disposal are required. In addition, the massive chemical dosages which are sometimes used in advanced wastewater treatment make chemical recovery and reuse attractive. Conditions are such that sludge dewatering will be required in most advanced wastewater treatment plants.

The second question to be explored in relation to sludge handling is why mixtures of chemical and sewage sludges are often very difficult to dewater—although this may not always be the case. There is some evidence to indicate that the chemical composition of the wastewater, particularly the hardness, has an important effect. In hard water areas, chemical-sewage sludge mixtures may dewater more readily than in extremely soft waters where the resulting mixtures may be extremely difficult to dewater. It has been suggested that this difference, if it in fact exists, may be related to different concentrations of sulfates, magnesium, total dissolved solids, or other substances in the wastewater. No specific reason for this apparent difference has been established definitely. If a chemical coagulant is used prior to primary clarification, there is one distinct difference in the settled sludge. It contains chemically coagulated colloids not normally present in primary sludge and more trapped water than primary sludge from plain sedimentation. The colloidal fines are more difficult to separate from the associated water than are the coarser primary solids. The *p*H of the sludge is an important factor. Generally, if the *p*H is below 7.0, the sludge is easier to dewater. Because there are many unknowns affecting dewatering of chemical-sewage sludge mixtures, it is a wise precaution to investigate thoroughly the dewatering characteristics of sludge mixtures from the treatment of a particular wastewater prior to final selection of a plant flow sheet and design of a full-scale plant. This investigation should be on a pilot plant scale, for present laboratory techniques do not adequately duplicate plant conditions.

There are two methods for pretreatment of chemical-sewage sludge mixtures which may aid greatly in final dewatering: anaerobic sludge digestion and heat treatment. Successful anaerobic digestion of sewage-chemical sludge mixtures often produces sludge which dewaters quite readily. Phosphorus which has been removed from wastewater by alum or lime coagulation remains with the sludge and is not present in appreciable quantities in the supernatant liquor. On the other hand, phosphorus removed biologically is released during anaerobic digestion and becomes dissolved in the supernatant liquor only to be recycled to the process which is undesirable. If digested sludge is to be hauled wet to disposal or dried on sand beds, then sludge digestion may be a complete answer to dewatering although removal of nutrients from the digester supernatant may be necessary to prevent recycling of nutrients to a biological process. Digested sludge must be dewatered further before it can be incinerated either for final disposal or for chemical recovery. Unfortunately, anaerobic sludge digestion is subject to the well-known difficulties of maintaining and controlling a biological process.

Another method of pretreatment of sludges which may lend itself to chemical-sewage sludges is heat treatment, such as the Porteous Process, which utilizes heat and pressure to break down the gelatinous structure of sludge with the release of the associated water. In this process, sludge is taken from a digestor or thickener to a grinder. Then the sludge is pumped at a pressure of about 250 psi through a heat exchanger and into a reaction vessel. The sludge temperature is brought up to about 350°F in the heat exchanger. Steam is injected directly into the reaction vessel to reach the operating temperature of 380°F. In addition to raising the temperature, steam injection has a beneficial effect on the dewatering characteristics of the sludge. Hot sludge is held in the reaction vessel for about 30 min, and is then passed back through the heat exchanger for heat recovery, and cooled. The sludge can be further dewatered to a solids content of 50–55 percent without the need for conditioning chemicals by means of a vacuum filter, centrifuge, or filter press.

Centrifuging or vacuum filtration, preceded by float thickening of waste activated sludge are the methods which probably will find the widest use in advanced wastewater treatment plants. Filter presses are another potential means, but they have not been widely used in the United States to date because of their high initial cost and extensive floor space requirements. Dewatering mixtures of primary and waste activated sludge by centrifuging requires the addition of a polymer at a cost of from $2–12 per ton (1970 prices) of dry solids. The use of a float thickener plus coil filter is competitive from a cost standpoint. Direct dewatering of certain chemical-sewage sludge mixtures by centrifuging or vacuum filtering may involve chemical conditioning costs as high as $50–60 per ton of dry solids, which is prohibitive. There are two things which can be done. One is to use pretreatment of the sludge mixture by digestion or heat treatment. The other is to settle and process the sewage and chemical sludges separately. The latter method has been successfully used in a full-scale plant at South Tahoe employing a concurrent flow centrifugal. It has also been pilot tested successfully at the same location using a coil filter.

SYSTEMS FOR HANDLING OF ORGANIC SLUDGES

The sludge handling system at the South Tahoe Water Reclamation Plant will be used to illustrate the satisfactory handling of mixtures of primary and waste activated sludge, and the problems encountered in attempting to dewater some chemical-sewage sludge mixtures.

Figure 9-1 is a schematic diagram of this equipment installation. Very briefly, the process consists of dewatering primary and waste activated sludge in concurrent flow centrifugals with addition of a polymer (at the rate of 1–10 lb/ton of dry solids), from an initial solids content of 1–2 percent to about 15–22 percent; incineration in a multiple hearth furnace; and disposal of the insoluble, sterile ash.

The incineration system is designed to handle sludge from a flow of 7.5 mgd of domestic wastewater. The anticipated amount of dry solids per day from this flow is as follows: raw primary sludge, 8,200 lb; waste activated sludge, 9,800 lb; and waste calcium sludges, 3,000 lb (from lime system); for a total of 21,000 lb. The fuel value of the sludges in Btu/lb of dry solids are raw primary 7,000; waste activated, 7,500; and calcium sludge, 0.

Two plunger-type sludge pumps equipped with variable speed drives are used to deliver sludge to the centrifugals. The maximum capacity of each sludge pump is 120 gpm.

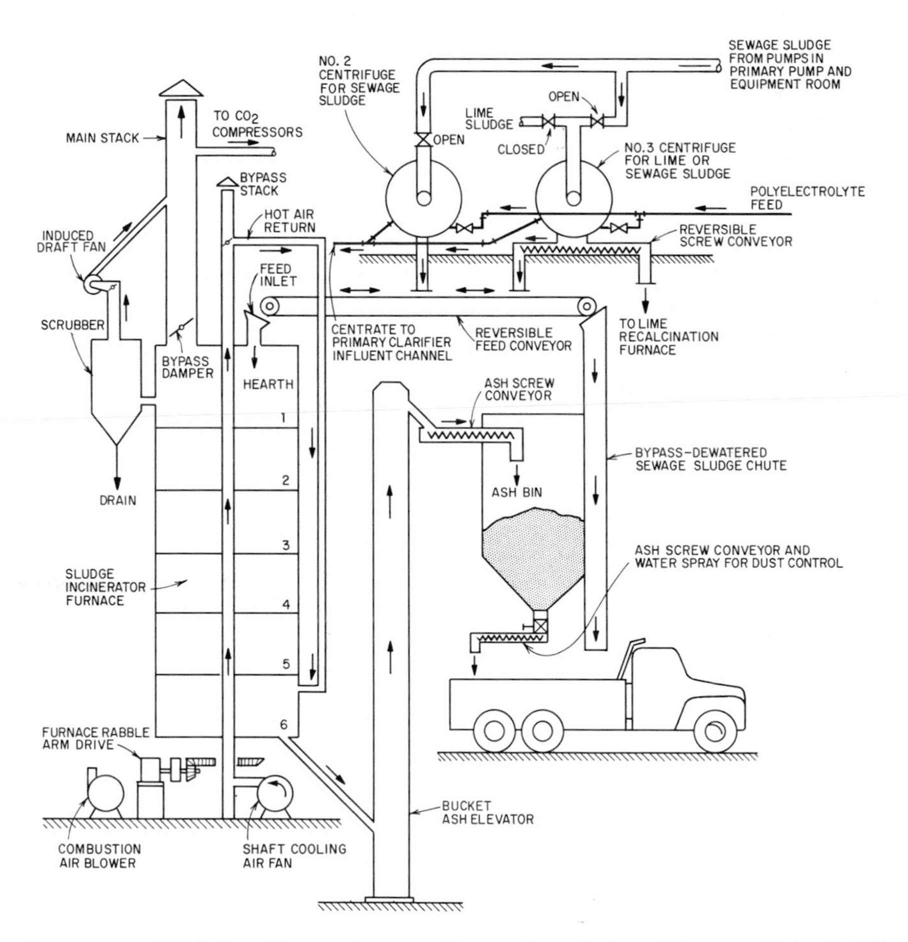

Figure 9–1 Solids handling—primary and secondary sludge. *(Courtesy Clair A. Hill & Assocs.)*

ORGANIC SLUDGE DEWATERING BY CENTRIFUGE

Two identical machines are installed for sewage sludge dewatering. They are Bird Machine Company 24 × 60 in. concurrent flow centrifugals as pictured in Figure 9-2. A solid bowl centrifugal is a separation device for the rapid and thorough separation of suspended solids from a liquid slurry by using centrifugal action to create separation forces on solid particles which are some thousand times that of gravity. The separation vessel, or bowl, rotates at high speed to create the settling forces. It has an overflow weir for collection and dis-

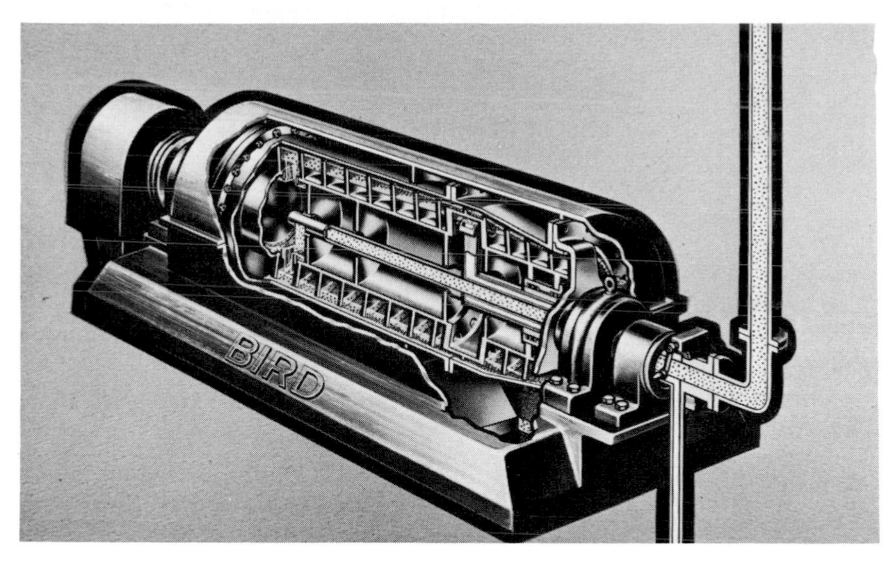

Figure 9–2 Bird concurrent flow centrifugal. *(Courtesy Bird Machine Co.)*

charge of the liquor or centrate, and a drainage deck equipped with a conveyor which continuously picks up and removes the settled solids. In dewatering sewage solids, it is important to obtain good clarity of the centrate, so as not to recycle excessive amounts of fines back through the process by way of the centrate. The factors influencing dewatering of waste sludges by centrifuging include feed rate, feed consistency, bowl speed, pool depth, conveyor speed, and chemical aids (usually polymer solutions). The clarity of the centrate may be improved by increase in bowl speed, increase in pool volume, increase in conveyor speed, reduction in consistency of feed slurry, reduction in feed rate, or increase in chemical treatment. It is also desirable to improve the dryness of the cake in order to reduce the amount of fuel required for incineration, if this can be done without sacrificing centrate clarity. Cake moisture can be reduced by increasing the bowl speed, decreasing the pool volume, or reducing the conveyor speed. Increased temperature benefits both centrate clarity and cake dryness. The best polymer for treating a particular sludge is a matter of trial and error. A cationic, an anionic, or both may be suitable depending on the type of sludge processed. At Tahoe the polymer feed system for the sludge centrifugals consists of a single mix and feed tank, and a Moyno pump equipped with variable speed drive. Positive means are provided for controlling the feed rate of sludge solids to the centrifugals. At Tahoe, the centrifuging of mixtures of primary and waste activated sludge, either raw or digested, has been quite satisfactory, although sludge mixtures pretreated by digestion dewater much more readily than do raw sludges. Machine torques and chemical costs are both lower with digested sludge mixtures. In either case, performance is good and costs are reasonable. On the other hand, in attempting to centrifuge mixtures of raw primary, waste activated, and lime sludge many operating problems were encountered and costs were high. These chemical-sewage sludge mixtures were very difficult to dewater even when large quantities of polymer were used. This is a property of the sludge rather than the equipment used, it appears, because similar difficulties were experienced in laboratory experiments and in pilot tests of vacuum filtration. Several months of extensive experimentation failed to reveal any satisfactory means of dewatering this particular chemical-sewage sludge mixture. Finally, the problem was solved by separate settling and processing of sewage and chemical sludges.

At Tahoe, the centrate from the sludge centrifugals returns to the primary tank. The sludge cake drops onto a conveyor belt which feeds the sludge to the incineration furnace. The belt may also be reversed for delivery of dewatered sludge to an outside chute for truck loading. The capacity of each 24″ × 60″ concurrent flow centrifugal is about 450 lb of dry solids per hour. The centrifugals operate best with a solids concentration of less than 2 percent in the feed of mixtures of raw primary and waste-activated sludge. The thin slurry seems to improve the conveyance of solids through the machines. Suspended solids capture is more than 90 percent with the proper addition of polymer.

Multiple Hearth Incineration of Organic Sludge

For sludge incineration, a multiple hearth furnace supplied by B-S-P Corporation is used. It is a 6-hearth unit with an inside diameter of 14 ft 3 in. It consists of a circular steel shell with six refractory hearths. There is a central rotating shaft with rabble arms attached. The shaft and arms are hollow and air is used to keep the assembly cool. The cooling air is either vented to the hot air stack, or recycled through the hot air return duct to the bottom hearth for preheating sludge combustion air. The shaft cooling air fan is connected to an emergency power source to protect the shaft and rabble arms in event of a power failure. The upper hearth is an in-feed hearth, and the second is an out-feed hearth, and so on. Sludge is rabbled into the center of hearth 1 and drops through a center opening to hearth 2 where it is rabbled to drop holes at the outside perimeter down to hearth 3, and so on. Burners are attached at certain locations in the shell to heat the furnace to the required temperature. The furnace has an induced draft fan which is operated to maintain a 4–6 in. W.C. differential pressure across the stack gas scrubber in order to insure good fly ash removal. In emergency power failure there is a main bypass damper which opens, venting the furnace gases to the atmosphere. The incinerator is also equipped with a combustion air fan which furnishes air for the pilot and main burners. There are eight nozzle-mix gas burners each with its own separate aspirating pilot burner, pilot solenoid valve, pilot atmospheric regulator, gas cock, air cock for pilot, burner nozzle, main regulator ratio adjust valve, safety shut-off valve, and main gas cock. For firing rate control there are four manual and four automatic butterfly air valves. The main gas supply is natural gas, and the standby supply is propane. Vaporizers and blendors evaporate and dilute the propane with air to match the composition of the natural gas supply in order to permit its use in the furnace burners without change in the firing system. A Sly Company scrubber is installed to thoroughly clean the exhaust gases before they are vented to the atmosphere. The gas is cooled to 110° by the scrubbers. There are also burners on the top furnace hearth which can be used as afterburners if necessary to prevent air pollution. There is no visible smoke or steam plume and no odor from the exhaust stack of the organic sludge incinerator.

Furnace feed mixtures of raw primary and waste activated sludge with solids content of 20 percent or more are autocombustible. Wetter mixtures of these sludges and digested sludges require supplemental fuel. Usual furnace temperatures are, in °F, hearth No. 1, 750; No. 2, 1400, No. 3 1600; No. 4, 1600; No. 5, 1500; No. 6, 500.

The sludge cake from the sludge dewatering centrifuges is conveyed to the incineration furnace by a belt conveyor and drops through a flap gate into the outside of the top hearth. The sterile insoluble ash, which resembles volcanic ash in appearance, falls through a chute into a bucket elevator which lifts the ash from the basement of the three-story building to near the ceiling, where it is conveyed laterally by means of a screw conveyor to an ash storage bin outside the building where the ash can be loaded into trucks to be hauled to land disposal. The ash is suitable for by-product use in making building bricks or in stabilizing highway and road subgrades (as a substitute for portland cement), but neither use is a profitable one at present. When the furnace is down for repairs or servicing,

Figure 9–3 Control panel in incineration building. *(Courtesy B-S-P Corp.)*

one of two things can be done. Sludge can be stored in old sludge digestion tanks or the sludge can be dewatered by the centrifuge and, by reversing the conveyor belt, be conveyed to a truck-loading chute for wet disposal.

The sludge incineration control and instrumentation system permits observation and control of many functions from a central panel (Figure 9-3). There are temperature devices to measure and record temperature at each hearth, cooling air exhaust, scrubber inlet gas, and ID fan inlet gas. There are draft gages for measuring flue gas pressures at various points. Automatic controls are provided to maintain the temperature on each fired hearth within plus or minus 40°F from the set point. Sensing devices are installed to monitor certain malfunctions which are then reported to a central annunciator panel. Alarms are provided for burner shutdown, furnace overtemperature, draft loss, feed belt shutdown, and ash disposal system failure. In the event of power or fuel failure, the furnace shuts down automatically until manually reset, and the shaft cooling air fan is automatically transferred to the standby power source.

Costs of Centrifugation and Multiple Hearth Incineration

At Tahoe for 7.5 mgd capacity, the capital costs for organic sludge handling, dewatering, incineration, and ash disposal are $545,000 at the 1969 FWPCA STP Construction Cost Index = 127.1. At design capacity, the operation costs per ton of dry solids are about $25 for handling and dewatering sludge, and $27 for sludge incineration and ash disposal. With 1.2 tons/mg of organic sludge, these costs are equivalent to $30/mg and $32.40/mg, respectively.

OTHER METHODS FOR ORGANIC SLUDGE CONDITIONING, DEWATERING, AND INCINERATION

Organic sludge can be conditioned by anaerobic digestion, elutriation, chemical treatment, heat treatment, float thickening, gravity thickening, wet oxidation, or the addition of recycled furnace ash or other solids.

Because they occupy much less space and are unaffected by weather conditions, the use of mechanical methods of dewatering sludges is increasing, while the use of sludge drying beds is decreasing. As mentioned previously, float thickening of waste activated sludge and dewatering of mixtures of this thickened sludge with raw primary sludge by a vacuum filter with metal coil or cloth septum is competitive in performance and cost with the centrifuging operation. Sludge dewatering may also be accomplished in a filter press. The sludge cake from any of these dewatering processes can be burned in a fluidized bed furnace as an alternate to incineration in a multiple hearth furnace.

Sludge Conditioning

Mention has already been made of the fact that anaerobic digestion of primary and secondary sludges breaks down water-binding organics so that the digested sludge dewaters much more readily. This also applies to mixtures of organic and chemical sludges. However, digested sludges often require higher chemical dosages for mechanical dewatering due to the increase in alkalinity which occurs during digestion. Elutriation may be used to reduce the alkalinity and thus the chemical demand of digested sludge. A 2:1 ratio of elution water to wet sludge is often used. The chemical demand may be reduced by as much as one third by elutriation of digested sludge. One serious disadvantage of the elutriation process is the fact that the fine solids which are washed out of the sludge are very difficult to coagulate and settle out when they are returned to the liquid treatment process. The advent of polyelectrolytes and their use in sludge conditioning has eliminated the need for elutriation because the action of polymers is not adversely affected by high alkalinity sludges.

Chemical conditioning of sludge may be accomplished by chlorine, alum, lime, ferric salts, or polymers. Under proper conditions, the use of chemical conditioners flocculates the sludge solids to form porous agglomerates which allow the water to be more easily separated from the solids. Coagulant usage can be significantly reduced by adequate mixing or stirring and by use of fresh sludge. Storage of raw sludge for any significant period of time may increase chemical requirements. Polymers are now used almost universally for sludge conditioning prior to dewatering. They are usually more effective, easier to handle, and more economical than other chemicals. Heavy chlorination, 20 lb of chlorine per 1,000 gal, of waste activated sludge permits gravity thickening from 0.5–10 percent solids. The required reaction time is only about 2 min. Usually about a pound of lime per pound of chlorine is added to avoid excessive *p*H depression. Tenney, *et al.* (1970) have shown that the chemical conditioner dose required for optimum dewatering of waste activated sludge is affected by the *p*H, alkalinity, and phosphate content of the wastewater, and the solids concentration, primary waste activated mix ratio, and storage time of the sludge. For the sludges which they tested, the optimum *p*H for effective conditioning was in the range of 6.0–7.0 for iron salts, and 4.5–5.5 for alum salts. They also report that increasing solids concentrations in sludge require proportionate increases in conditioner dose. They found that increasing orthophosphate concentrations in sludges also require larger dosages of conditioning chemicals. They concluded that the reason that sludge holding increases chemical requirements is that a digestive breakdown occurs which produces a greater number of small sludge particles which in turn exert a higher conditioner demand.

Heat treatment by the Porteous Process as a method of sludge conditioning has already been discussed earlier in this chapter. Another method for heat treatment is wet oxidation by the Zimpro Process. Blattler (1970) in reporting on wet air oxidation at Levittown describes his experiences with this process for treatment of raw or digested sludges. Sludge to be processed is ground to particles less than 0.25 in., and then pumped at 300 psi to the process. Pressurized air is added, and the mixture then enters a heat exchanger where the temperature is raised to 300°F by the main reactor effluent. In the reactor, the oxidation reaction takes place. Steam is added to obtain a reactor outlet temperature of about 350°F. Treated sludge and exhaust air leaving the reactor are cooled by the incoming sludge and air as the mixtures flow countercurrently through the heat exchanger. The off gases are passed through a catalytic deodorizer. Oxidized sludge passes from a sludge-gas separator to a gravity thickener. Thickened sludge is either wet-hauled or vacuum filtered. Chemicals are not required for vacuum filtration of the oxidized sludge. Blattler (1970) reports a filtration rate of 7 lb/ft^2/hr, and a suspended solids capture of 96.5 percent. The solids content of the filter cake was 40–45 percent. Sludge dewatering of the oxidized sludge by centrifuge was carried out experimentally. Centrifuge cake solids averaged about 35 percent and suspended solids capture about 80 percent. No odor was observed in the cooled liquid, and no bad effects of filtrate return to the plant were noted. In other locations using this process, reaeration of the filtrate return has been practiced to avoid overloading the biological processes.

Furnace ash from the incineration of organic sludges is an excellent sludge conditioner. Recycled ash is often used as a conditioner in sludge dewatering by filter press. At South Tahoe, unusual plant-scale experiments were conducted in which recycled furnace ash was added to flotation thickened mixtures of raw primary, waste activated, and waste lime sludges and then fed directly into a multiple hearth furnace and incinerated without the use of any mechanical dewatering device. The sludge mixture was thickened to about 10 percent solids in the flotation thickener, mixed with an equal volume of recycled ash in an elevating screw conveyor, and then fed directly into the furnace. The favorable results indicate that further investigation of this method is warranted. Table 9–1 lists the results of laboratory work at Tahoe

Table 9-1 Sludge Conditioning with Recycled Furnace Ash.

Sludge Type	*Weight of Ash (g)*[1]	*Time to Dewater (sec)*	*Filtrate Volume (ml)*
Primary-	1	155	95
Calcium	3	30	92
Mixture	5	16	90
	7	13	89
Waste	3	722	92
Activated	5	140	90
	7	55	89
	10	30	86

[1]The sludge volume used was 100 ml in all cases. The ash contained 18.4 percent CaO from the waste calcium sludges. The sieve analysis of the ash is given below:

Percent Passing by Dry Weight

Screen Opening	3/8″	#4	#8	#16	#30	#50	#100	#200
	100	99	99	96	80	59	32	8

which illustrates the improvement in dewatering rates of sludge from the addition of furnace ash.

In other places other inert materials such as ground oyster shell, fly ash, decomposed granite, and magnesite have been used to thicken waste activated sludge to as high as 25 percent solids.

Flotation Thickening

Waste activated sludge is usually about 99.5 percent water which is very difficult to separate from the associated solids. It is often helpful or necessary to thicken waste activated sludge to about 5 percent solids to prepare it for more complete dewatering in a vacuum filter or other device. Float thickeners have been used very successfully to thicken waste activated sludge to 4–6 percent solids. One such unit is the H-R Flotation Thickener as manufactured by Komline-Sanderson Engineering Corporation of Peapack, N.J. Figure 9–4 is a schematic flow diagram of this unit and its auxiliaries. The thickener consists of the main flotation basin equipped with a sludge removal mechanism at the water surface, a recirculation pump, a compressed air supply, an air dissolution tank equipped with a reaeration pump, and a polymer feed system. The thickener operates as a dissolved-air-type flotation unit. The minute bubbles necessary for flotation of the sludge are produced by dissolving air in the recycled effluent at 50–70 psi. The recycled flow is mixed with the basin influent at reduced pressure. The tiny bubbles of air expand and attach to sludge particles and cause them to float to the top water surface where the sludge par-

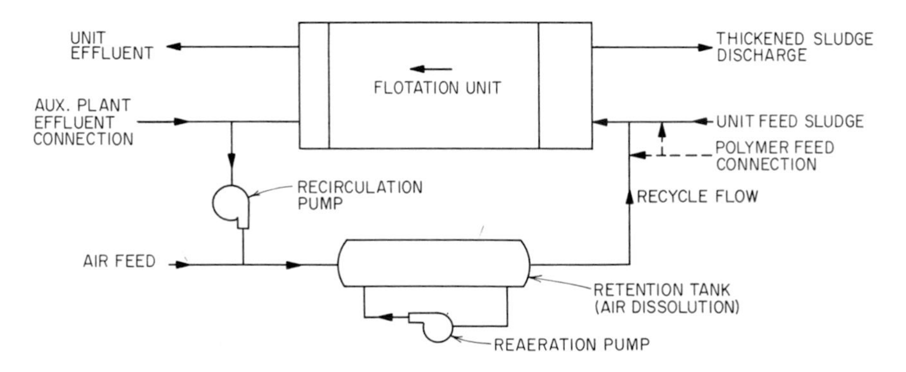

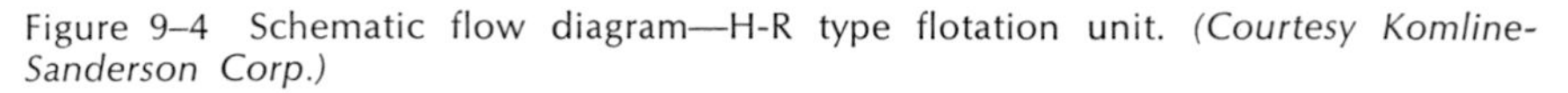
Figure 9–4 Schematic flow diagram—H-R type flotation unit. *(Courtesy Komline-Sanderson Corp.)*

ticles agglomerate and condense to form a thick sludge blanket. The sludge blanket is continuously removed from the water surface by the mechanical collector. High molecular weight organic polyelectrolytes are excellent flotation aids when added to the influent to the thickener unit. Design loadings for flotation units are about 2 lb dry solids per square foot per hour, and 0.8 gpm/ft^2 hydraulic loading. If there is an overload of solids, there will not be enough air to float the sludge. A hydraulic overload will result in scouring of solids off the bottom of the sludge blanket by the excess flow at high velocity of water beneath.

Gravity Thickening

Gravity thickening of waste activated sludge or primary sludge or mixtures has been practiced for the past 40 or 50 years. Mechanical stirring is often used to assist the gravity thickening process. Several factors influence the extent to which sludge can be thickened, including the type of sludge, initial solids concentration, settling time, temperature, use of chemicals, and operating conditions. A low SVI (sludge volume index) contributes to good thickening. Primary sludge is more readily thickened than waste activated sludge. Hydraulic surface loading rates of about 100 gpd/ft^2 are commonly used for thickening. Typical solids loading for thickeners in lb/day/ft^2 are about 5 for waste activated sludge, 10 for waste activated-primary sludge mixtures, 10 for trickling filter sludge, 10 for trickling filter-primary sludge mixtures, and 25 for raw primary sludge. Typical thickened sludge solids content in percentages are about 3 for waste activated sludge, 6 for mixtures of waste activated and primary, 8 for trickling filter sludge, 8 for trickling filter-primary mixtures, and 10 for raw primary sludge. The volume of sludge is reduced to about 20 percent of the original sludge volume. The rate of thickening is higher at elevated water temperatures. The addition of lime, chlorine, or a polymer assists in thickening. It is not practical to thicken sludges to more than about 10 percent solids by gravity thickening.

Vacuum Filtration

Vacuum filtration and centrifuging are alternate satisfactory means for dewatering raw primary and waste activated sludge mixtures. The ability of vacuum filters to dewater sewage-chemical sludges has not been fully explored, and plant-scale tests are needed to determine their capability in this regard.

A vacuum filter installation includes a sludge feed pump, chemical feeders, sludge conditioning tanks, vacuum filter, vacuum pump and receiving tank, filtrate pump, and filter cake conveyor and hopper. The vacuum filter consists of a suspended drum which rotates in a container of sludge with about one fourth of the outer surface of the drum being submerged. Sections of the drum are alternately subjected to suction and pressure during each revolution. The surface of the drum is covered with a filter medium through which the water is drawn as the drum passes through the sludge. The solids are deposited as a cake about one-quarter inch thick on the filter medium. As this cake rises out of the liquid sludge, air is drawn through the cake for further dewatering. The cake is then scraped off the filter medium before it is again submerged to pick up a new layer of sludge. The filtrate is returned to the treatment process. Vacuum filters usually will produce a cake which contains about 20–40 percent solids at a rate of approximately 5 lb of dry solids per square foot per hour. Dewatering is greatly assisted by use of sludge conditioning chemicals such as polymers, or lime, or, if absolutely necessary, ferric chloride.

The filter medium used has an important bearing on vacuum filter performance. The two general types are coil springs and cloth. The Komline-Sanderson Coilfilter uses two layers of coil springs. The filtrate passes through and between the coils. As the drum revolves, one layer of the coils separates from the other and conveys the dry sludge cake to a point of discharge from the machine. The coils are then washed before making another revolution around the drum. There are a number of vacuum filters which use cloth or fabric media. Wool, cotton, Orlon, nylon, Dacron, polyester fiber, and other materials are used. Vacuum filters with cloth or fabric medium are supplied by Komline-Sanderson, Dorr-Oliver, Walker Process, Eimco, Rex Chainbelt, and other companies.

Sludge should be concentrated as much as possible before application to a vacuum filter. This increases filter yield and reduces the volume of filtrate that must be recycled. An almost linear relationship exists between feed concentration and filter yield. At 6 percent solids the filter rate is about 3 $lb/ft^2/hr$, while at 10 percent solids the filter rate is about 6 $lb/ft^2/hr$. Filter cake solids increase from about 30 percent with application of sludge with 6 percent solids to about 35 percent at 10 percent solids in the feed to the filter.

Heat treated sludge dewaters quite well even at high rates on the vacuum. Forty percent solids can be obtained at rates of 7 $lb/ft^2/hr$ and temperature of 380°F. on most types of organic sludges.

Countercurrent Centrifugals

Concurrent flow centrifugals were discussed previously in describing operations at Tahoe. The concurrent flow centrifugal is designed to make it possible to settle extremely fine suspensions. The slurry is fed into the bowl ahead of the settling zone. Previously settled solids are not disturbed by incoming feed or by the turbulence of conventional countercurrent machines. Conveying solids over the entire length of the machine is said to produce better compaction of solids and a drier cake.

However, countercurrent machines have been used successfully in many wastewater treatment plants, and probably will continue to find future application in the field.

Filter Presses

Filter presses have not been widely used in the United States because of their high initial cost and their large floor space requirements as compared to other methods of mechanical dewatering. Former high labor costs have been overcome by almost complete automation of filter press systems.

The principal advantages of the filter press are (1) the high cake

Figure 9–5 A vacuum filter installation in operation. Note the sludge cake dropping from the filter. *(Courtesy Komline-Sanderson Corp.)*

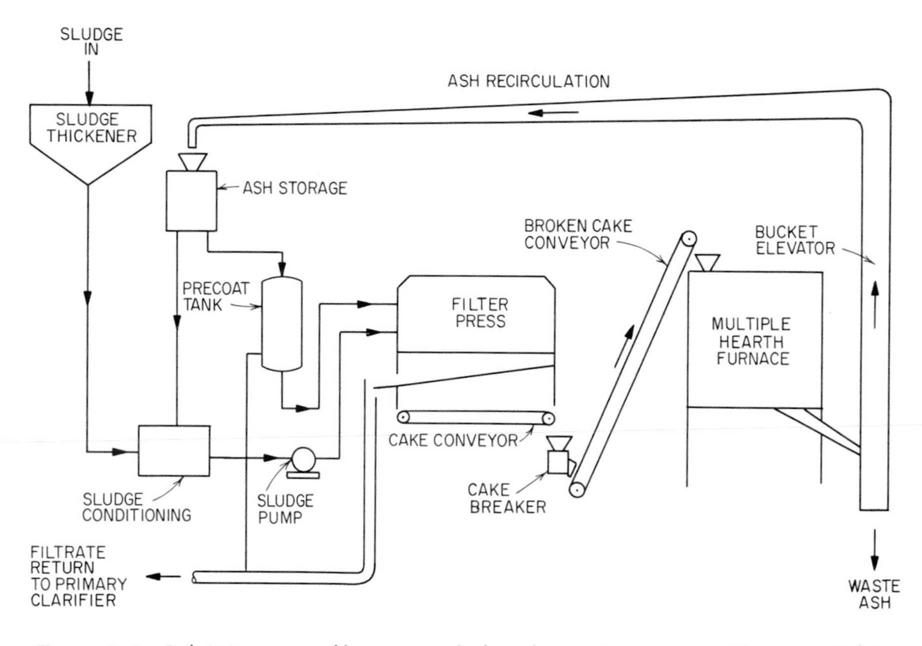

Figure 9–6 Beloit-Passavant filter press sludge dewatering system. *(Courtesy Beloit-Passavant Co.)*

solids produced, about 40–45 percent; (2) its capability to handle any type of sludge, including sewage-chemical sludge mixtures; (3) the clear filtrate produced, which contains only 10–20 mg/l of suspended solids and less than 200 mg/l of BOD; and (4) the elimination of the need for chemical conditioning agents.

A typical layout for a Beloit-Passavant "Sludge-All" single step dewatering system is given in Figure 9–6.

The operation of the filter press system is as follows: sludge is pumped to a gravity thickener. The overflow from the thickener usually is returned to the primary clarifier. Thickened sludge is pumped to a sludge grinder to reduce the particles to a size which will prevent clogging or damaging the equipment which follows in the process and to make the sludge more amenable to dewatering. Ash is added as a conditioning agent and mixed with the sludge. The filter press is precoated with an ash-water slurry. After this is done, the sludge is then pumped to the filter under pressure. As dewatering progresses in the filter, more sludge is pumped to the press to maintain a constant pressure of 200 psi within the press. The total cycle time for dewatering is about 1½ hr. At the end of the dewatering cycle, the filter press is opened and the sludge cake drops from the dewatering chambers onto a conveyor. The cake is then passed through a breaker or grinder and is then conveyed to the incinerator. The ability of the press to produce a cake with 40–45 percent solids instead of about 20 percent as obtained by other dewatering methods reduces fuel consumption in the incineration process. In cases where sludge incineration is not practiced, fly ash from other sources may be used in place of the furnace ash for sludge conditioning.

Fluidized Sand Bed Furnaces

An alternate to multiple hearth furnaces for incineration of sewage sludge or recalcining of lime is the fluidized sand bed reactor, as illustrated by the Fluo-Solids System manufactured by Dorr-Oliver, Inc. (Figure 9–7). Sludge is prepared for the fluidized sand bed reactor by centrifuging or vacuum filtration. Within the reactor, sand is suspended by air at a temperature of about 1500°F. The solids are dispersed throughout the bed and burned to ash. The off gases are water scrubbed before release to the atmosphere. Inert solids are removed from the scrubber flow by a cyclone separator.

SYSTEMS FOR HANDLING OF SLUDGE RESULTING FROM CHEMICAL COAGULATION

Lime Recalcining and Reuse. General Considerations

Lime recalcination is an old art in many industrial applications. It is also used to some extent in water works practice, principally in lime-soda water softening plants. The recalcining and reuse of lime in wastewater treatment plants is new, but it is unquestionably destined to see a rapid increase in the number of applications in this field. Phosphorus removal by lime coagulation requires massive doses, in the range of 100–500 mg/l as CaO, of lime. This high rate of lime addition produces large quantities of sludge. Roughly, the volume of lime sludge produced is about 1⅓ times that of organic sludges resulting from secondary treatment, which gives some perspective to the magnitude of the problem. If the lime sludge emanating from a wastewater treatment plant can be disposed of satisfactorily on land, then this may be the cheapest arrangement. In the absence of this expedient for lime sludge disposal, there is a real problem. One obvious solution is to recalcine and reuse the lime within the plant. The re-

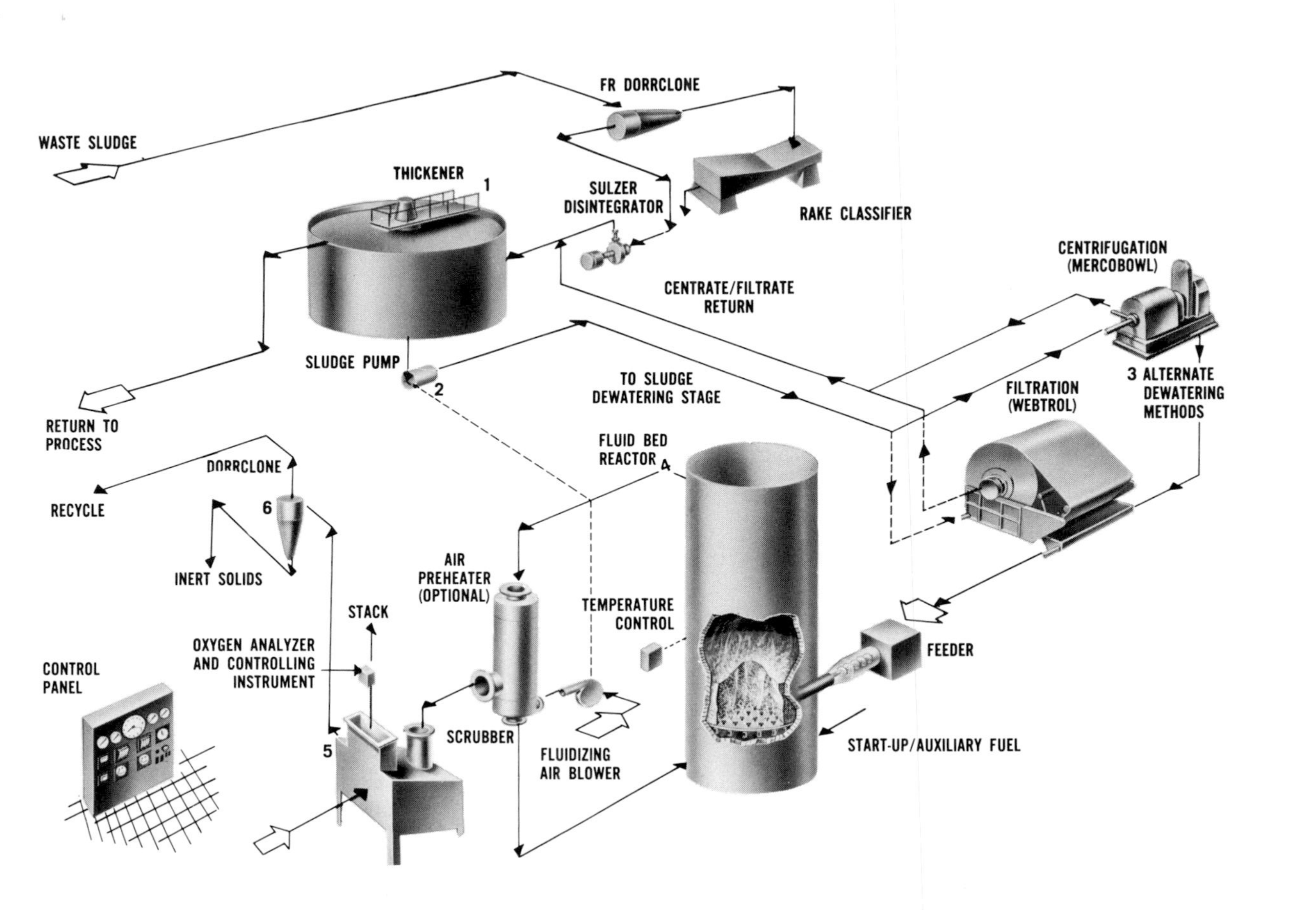

Figure 9–7 Dorr-Oliver FS disposal-system flow sheet. *(Courtesy Dorr-Oliver, Inc.)*

calcining process is quite simple. It consists of heating the calcium sludge to temperatures of about 1850°F which drives off water and carbon dioxide leaving only the calcium oxide or quicklime. In all cases there will still be some sludge to be disposed of but the total volume will be greatly reduced. Some lime sludge must be wasted periodically to avoid a buildup of inerts. The amount to be wasted is an economic balance between solids handling capacity and desired characteristics of the reclaimed lime. In soft water areas, some make-up lime will be required to replace the calcium which leaves the plant dissolved in the effluent water. In treating wastewaters high in calcium hardness an excess of recalcined lime may be produced. In small wastewater treatment plants (say 5 to 10 mgd), the cost of reclaimed lime may be only slightly less than the cost of new lime, but recalcining still may be warranted because it also saves the cost of disposing of the lime sludge. In large plants (over 10 mgd), recalcining probably can be justified by the lower cost of the reclaimed lime as compared to that for new lime.

At the South Tahoe Water Reclamation Plant, lime is used for clarification, phosphorus removal, and *p*H elevation for ammonia stripping. The spent lime sludge has been recalcined and reused in the process continuously for a period of more than two years. During that time more than 2,000 tons of recalcined lime have been produced. The system at Tahoe is typical of facilities for use in wastewater treatment plants, and will be used to illustrate process and design principles.

Tahoe Lime Recovery System

The lime system at Tahoe is designed for a plant wastewater flow of 7.5 mgd, with average lime dosages of 400 mg/l. Very briefly, the system consists of lime mud thickening to a solids content of 8–20 percent, further dewatering in a concurrent flow solid bowl centrifuge to 30–60 percent solids, recalcining in a multiple hearth furnace at temperatures of 1850°F, and return to reuse in the process. About 25–35 percent makeup lime is required. Figure 9–8 is schematic representation of the solids handling system for lime.

Lime Sludge Volume

If it is assumed that the underflow from the lime clarifier is 1 percent solids, then at 7.5 mgd flow and a lime dosage of 400 mg/l about 2,500,000 lb or 290,000 gal of liquid sludge per day is expected. This is about 200 gpm. If the solids content of the lime sludge is only 0.5 percent under the worst conditions, then the maximum flow might be

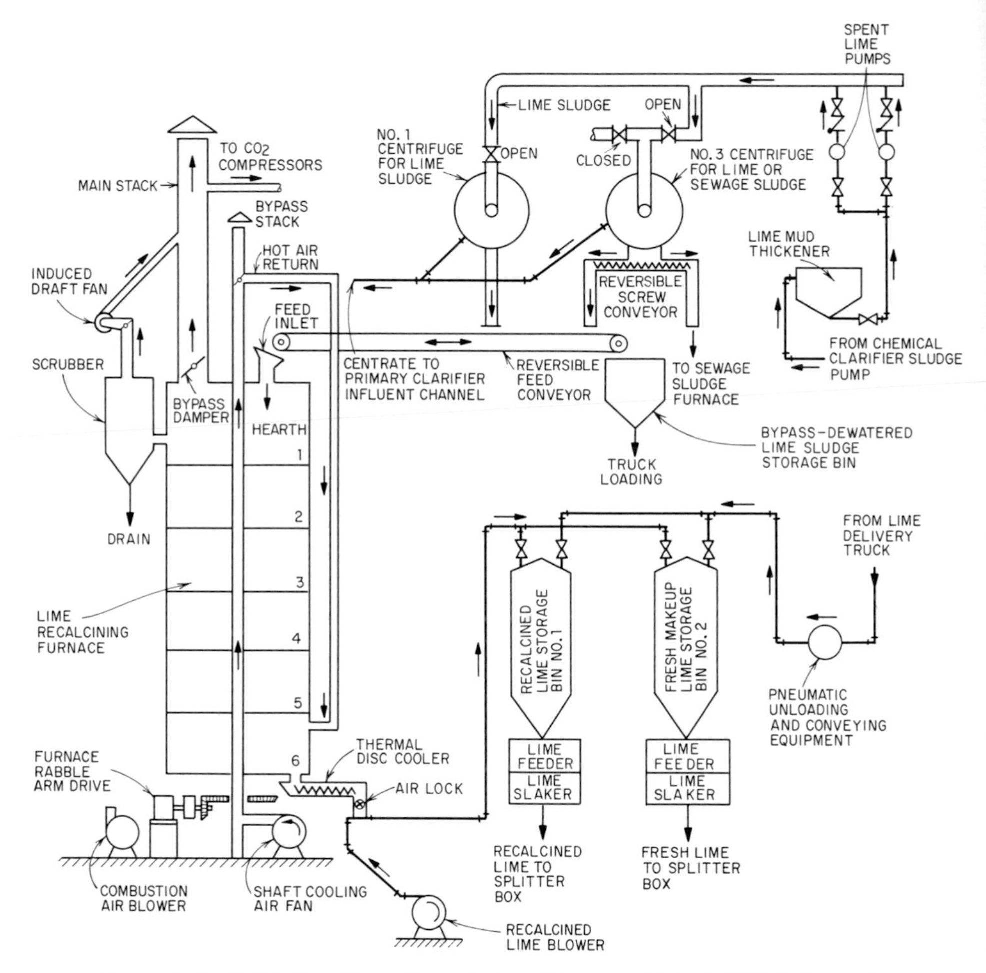

Figure 9–8 Solids handling—lime sludge. *(Courtesy Clair A. Hill & Assocs.)*

400 gpm. A single variable-speed horizontal centrifugal pump with a capacity of 450 gpm is installed at Tahoe for this service. The suction line under the floor of the chemical clarifier is glass lined. The discharge line is plain cast iron, and is cleaned periodically using a polyurethane "pig," forced through the line by water pressure applied behind it.

Lime Mud Thickener

The lime mud thickener is designed for a dry solids loading of not more than 200 lb/day/ft^2, and a surface overflow rate of 1,000 gpsf/

day. The unit installed is 30 ft in diameter by 8 ft deep with a Dorr-Oliver thickener mechanism of the bottom scraper type with variable speed drive. The thickener will allow withdrawal of lime sludge up to a solid content of about 20 percent, which has about the same consistency as toothpaste. Lime sludge thicker than this is very difficult to pump. The suction line to the pump for thickened sludge is glass lined, and is extended to the surface of the ground on the side opposite the pump suction line to permit rodding of the line beneath the bottom of the thickener basin. There are two Wemco 3-in., Model C, recessed impeller, horizontal centrifugal pumps with variable speed drives for delivering sludge from the thickener to the centrifuges. Each of these pumps has a maximum capacity of 40 gpm. Ordinarily sludge is withdrawn from the lime mud thickener at a consistency of about 8 percent solids, but under test the pumps have handled 20 percent lime solids satisfactorily.

Dewatering Lime Sludge

The thickened lime mud is further dewatered in a Bird Machine Company 24 × 60 in. concurrent flow centrifugal. This machine is identical to the sewage sludge centrifugals which were described earlier in this chapter. The maximum hydraulic capacity is 100 gpm. The maximum expected total solids feed to the lime centrifugal is 1,650 lb/hr. The minimum solids concentration in the feed is 8 and the maximum 20 percent. The maximum expected flow rate of feed to the machine is 36 gpm. In dewatering lime sludge, the centrifugal can produce a cake with 50–55 percent solids and can capture more than 90 percent of the solids. However, at Tahoe the lime centrifugal is used to classify or separate the phosphorus solids from the calcium solids, so that the phosphorus leaves the machine in the centrate, while the calcium is discharged in the cake. This is possible because of the greater specific gravity of the calcium solids. This type of classifying operation works very well, and is a good way to rid the lime recirculation system of the phosphorus. However, it has two disadvantages: it produces a wetter cake (about 40 percent solids) and a cloudier centrate than if the centrifugal were operated for best dewatering. Ideally there should be two centrifugals in series. The first would be operated to classify with the calcium cake going to the lime furnace with the phosphorus laden centrate going to the second centrifugal in the series. The second machine would be set to produce a centrate of high clarity, and the phosphorus cake would go to the sewage sludge incinerator. This type of series operation has been used on a trial basis at Tahoe and works exceptionally well.

The dewatered lime sludge is fed to the lime recalcining furnace by a belt conveyor. By reversing the conveyor, dewatered lime can be delivered to a truck loading bin outside the building for wet disposal.

Lime Recalcining Furnace

The lime recalciner is a 14 ft, 3 in. diameter 6-hearth B-S-P furnace. The furnace proper is identical to that previously described as used for sludge incineration. It is operated at higher temperatures, and there are different auxiliaries. The hearth temperatures are as follows in °F: hearth No. 1, 800; No. 2, 1250; No. 3, 1850; No. 4, 1850; No. 5, 1850; and No. 6, 750. It is important not to let the temperature drop too low on the top hearth in order to avoid clinker formation due to slow drying and balling of the lime cake. The lime furnace will deliver a maximum of 20,400 lb/day of recalcined lime plus impurities. The stack gas is wet scrubbed in a Sly Company jet impingement unit. There is no smoke or steam plume and no odor in the off gases.

At the bottom outlet from the hearth 6 of the lime furnace there is a lime grinder installed to break up large pieces of recalcined lime. It is protected by a coarse bar screen at the inlet. Following the grinder, the reclaimed lime passes through a termal disk cooler (water cooled). The lime is then conveyed to a recalcined lime storage bin of 35-ton capacity by a pneumatic system. This equipment consists of a stainless steel air lock discharging device, air compressor, and other accessories. It has a rated capacity of 0.75 tons/hr. There is also a pneumatic conveying system with dust collector for new makeup lime, and a second 35-ton storage bin for makeup lime.

Lime Feeding and Slaking

The makeup lime storage bin and the recalcined lime storage bin each discharge through gravimetric lime feeders and slakers. The feeders each have a capacity of 1,500 lb of CaO per hour. The lime slakers are Wallace & Tiernan paste-type slakers. The chemical feed system is discussed in more detail in Chapter 2.

Control and Instrumentation

The furnace controls are similar to those previously described for the organic sludge incinerator. In addition, the lime feeders may be selectively controlled either manually or automatically using recording *p*H meters at the point of lime application.

When the CaO content of the recalcined lime becomes less than 70 percent some dewatered lime sludge is wasted and the proportion of new makeup lime to recalcined lime fed to the wastewater is increased.

Costs

The capital costs for the complete lime recalcinating system for a plant capacity of 7.5 mgd and a lime dosage of 400 mg/l is $516,000 based on the 1969 FWPCA STP Construction Cost Index of 127.1. This includes lime mud handling, thickening, dewatering, and recalcining. At plant capacity, the operation costs are about $1.90/ton for lime sludge dewatering, and about $18.10/ton for recalcining, or a total operation cost of $20.00/ton of recalcined lime.

Other Lime Recalcining Furnaces

In addition to multiple hearth furnaces, fluidized sand bed furnaces, rotary kilns, or flash drying and recalcining systems are used in industrial and water works applications for lime recovery. These furnaces can also be applied to lime reclamation in wastewater treatment plants. The use of rotary kilns will probably be limited to large-capacity installations. The C-E Raymond System as built by Combustion Engineering, Inc., utilizes flash drying and calcination, and has been used in water softening plants for lime recovery.

ALUM SLUDGES

General Considerations

Alum is a good coagulant for many wastewaters. It is useful not only for clarification, but also for phosphorus removal. The alum dose for equal phosphorus removal will usually be about one half that required with the use of lime, although this must be determined for each individual wastewater. Alum costs about twice as much as lime per pound, so that chemical costs for phosphorus removal are about equal with the use of either lime or alum. Alum sludges may be disposed of satisfactorily in anaerobic digestion tanks along with sewage sludges. The phosphorus remains with the sludge and is not returned in the supernatant liquor. Alum sludges and alum-sewage sludge mixtures are exceedingly difficult to dewater. There are two basic methods for recovery of aluminum from settled alum sludges—the alkaline method and the acid method. Under circumstances which favor the use of alum over lime, these will be of interest.

Alkaline Method of Alum Recovery

Aluminum hydroxide is an amphoteric substance which can be dissolved in either an acidic or a basic solution. Lea, *et al.* (1954) suggested an alkaline method for alum recovery. The aluminum hydroxide is dissolved by raising the *p*H of the alum sludge to 11.9 with sodium hydroxide. The reaction converts the aluminum hydroxide to sodium aluminate and also returns the phosphates into solution. Calcium chloride is then added to react with the phosphates to form insoluble calcium phosphate. The reactions are as follows:

$$Al(OH)_3 \cdot PO_4 + 4NaOH \xrightarrow{pH11.9} Na_3PO_4 + 2H_2O + 3OH + NaAlO_2$$
$$NaAlO_2 + 2Na_3PO_4 + 3\ CaCl_2 \longrightarrow NaAlO_2 + Ca_3(PO_4)_2 \downarrow + 6NaCl$$

Lea, *et al.*, found in pilot studies that the reclaimed sodium aluminate and fresh alum were equally effective in removing phosphates from sewage. Contrary to these findings, several investigators have found sodium aluminate to be ineffective for wastewater coagulation. Sawyer (1952) found that sodium aluminate was unsuitable for phosphate removal on Boston sewage unless supplemental caustic or alum or sulfuric acid were added to respectively raise or lower the *p*H into a favorable range. Sawyer also found that in the *p*H range of 10–11, considerably more sodium aluminate was required when alum alone was used. In addition, Sawyer found that sodium aluminate was not effective in removing the complex polyphosphates at high *p*H levels. Slechta and Culp (1967) found that sodium aluminate was not an effective coagulant for coagulation of wastewater at South Lake Tahoe. Rose (1968) also reports that sodium aluminate was not as effective as alum in wastewater coagulation.

The first step in the evaluation of the feasibility of alkaline alum recovery is to determine the effectiveness of sodium aluminate as a coagulant by jar tests with the particular wastewater involved. If the aluminate appears to be a practical coagulant, then alkaline alum recovery should be considered. The following laboratory techniques will enable evaluation of the feasibility of alkaline alum recovery:

1. Coagulate 5 gal of wastewater with appropriate alum dosage.

Flocculate and settle. Decant clear liquid and divide sludge into 200-ml samples.
2. Treat sludge samples with varying quantities of sodium hydroxide to obtain *p*H values of 10–12.5. After the sludge is dissolved, determine the aluminum and phosphate concentrations in the resulting solution.
3. Add the stoichiometric amount of calcium chloride required to precipitate the phosphate.
4. Allow the calcium phosphate precipitate to settle. Decant the clear liquid, filter the remaining sludge, and combine the filtrate with the decanted liquid. Measure the aluminum and phosphate concentration in the liquid and the volume of recovered liquid. Calculate the percentage of alum recovery and cost of chemical required to recover alum.

The *p*H of the alum sludge must be raised to 12–12.5 with sodium hydroxide to maximize alum recovery. Alum recovery of 90–95 percent has been reported in the literature. Although the aluminum hydroxide is converted to soluble sodium aluminate at *p*H values of 10–10.5, a side reaction occurs when the calcium chloride is added which results in the precipitation of a substantial portion of the aluminum ion unless the *p*H is above 12.

The studies by Slechta and Culp (1967) showed, for the wastewater studied, that the chemical cost for alum recovery was about $62/ton recovered while the cost of fresh alum was $57/ton. In addition to the fact that the recovered alum was a poor coagulant, the fact that the cost of the recovered alum exceeded that of fresh eliminated alkaline alum recovery with sodium hydroxide and calcium chloride as a realistic alternate for the South Lake Tahoe plant.

An alternate alkaline recovery scheme uses lime as a source of the hydroxyl and calcium ions required to dissolve the aluminum hydroxide and precipitate the phosphate. However, alum recovery is usually substantially lower when using only lime. Maximum recoveries of about 35 percent have been reported.

Acidic Method

Aluminum hydroxide can also be dissolved by decreasing the *p*H by the addition of sulfuric acid to the alum sludge. Roberts and Roddy (1960) have reported on the successful recovery and reuse of alum at the Tampa, Florida, water treatment plant. The alum sludge from the clarifier is thickened and conveyed to a reactor where sul-

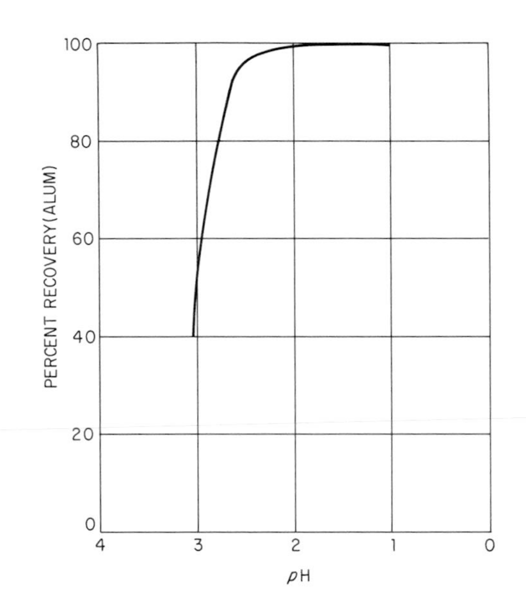

Figure 9–9 Acid recovery of alum. *(From Slectha and Culp, 1967)*

furic acid is added. The resulting alum solution is used as a coagulant for the raw water.

Slechta and Culp (1967) added sulfuric acid to the alum sludge resulting from alum coagulation of secondary effluent to obtain final *p*H values of 1–3. The resulting alum recovery is shown in Figure 9–9, and is in agreement with theory which indicates that aluminum hydroxide completely dissolves between *p*h 2.5–3. The acid-recovered alum proved to be an effective coagulant and it was estimated that its cost would be about one third the cost of fresh alum. However, the acid recovery also dissolves the phosphates which, if left in the alum solution, would be recycled to the effluent. If phosphate removal is the goal of alum addition, then such recycle of the phosphate would be intolerable. If coagulation for clarification only is the goal of alum addition, then acid recovery offers an attractive alternate for consideration.

Slechta and Culp (1967) also investigated methods of removing the phosphates from the acid-recovered alum. Use of anion exchange resigns and activated alumina were investigated but found to be impractical due to prohibitively high costs. An economical method of removing phosphates from acidified alum sludges has yet to be developed.

References

1. Albertson, O. E., and Guidi, E. E., Jr., "Centrifugation of Waste Sludges," *Journal Water Pollution Control Federation,* 1969, p. 607.
2. Alford, J. M., "Sludge Disposal Experiences at North Little Rock, Arkansas," *Journal Water Pollution Control Federation,* 1969, p. 175.
3. Aultman, W. W., "Reclamation and Reuse of Lime in Water Softening," *Journal American Water Works Association,* 1969, p. 640.
4. Bird Machine Co., South Walpole, Mass., *Operating Manual* (1966).
5. Black, A. P., and Eidsness, F. A., "Carbonation of Water Softening Plant Sludge," *Journal American Water Works Assoc.,* 1957, p. 1343.
6. Blattler, P. X., "Wet Air Oxidation at Levittown," *Water and Sewage Works,* Feb., 1970, p. 32.
7. B-S-P Corp., "Advanced Waste Treatment Seminar II," 1969.
8. Cardwell, E. C., "Dewatering by Mechanical Means," *Proceedings Tenth Sanitary Engineering Conference,* Univ. of Illinois Bulletin 65, 115:77 (1968).
9. Crow, W. B., "Techniques and Economics of Calcining Softening Sludges-Calcination Techniques," *Journal American Water Works Assoc.,* p. 322, (1960).
10. Ettelt, G. A., and Kennedy, T. J., "Research and Operational Experience in Sludge Dewatering at Chicago," *Journal Water Pollution Control Federation,* 1966, p. 248.
11. Genter, A. L., "Computing Coagulant Requirements in Sludge Conditioning," *Transactions, American Society of Civil Engineers,* 1946, p. 641.
12. Issac, B. C. G., and Vahidi, I., "The Recovery of Alum Sludge," *Proceedings Society Water Treatment and Examination,* 10:1 (1961).
13. Jones, W. H., "Sizing and Application of Dissolved Air Flotation Thickeners," *Water and Sewage Works,* Reference No. R–177–194 (Nov., 1968).
14. Jordan, V. J., and Scherer, C. H., "Gravity Thickening Techniques at a Water Reclamation Plant," *Journal Water Pollution Control Federation,* 1970, p. 180.
15. LaMer, V. K., and Smellie, R. H., Jr., "Flocculation, Subsidence, and Filtration of Phosphate Slimes. I. General," *Journal Colloid Science,* 1956, p. 704.
16. Lea, W. L., Rohlich, G. A., and Katz, W J., "Removal of Phosphates from Treated Sewage," *Sewage and Industrial Wastes,* 1954, p. 261.
17. McCarty, P. L., "Sludge Concentration-Needs, Accomplishments, and Future Goals," *Journal Water Pollution Control Federation,* 1966, p. 493.
18. Mulbarger, M. C., *et al.,* "Lime Clarification, Recovery, Reuse, and Sludge Dewatering Characteristics," *Journal Water Pollution Control Federation,* 1969, p. 2070.
19. Nelsen, F. G., "Recalcination of Water Softening Plant Sludge," *Journal American Water Works Assoc.,* 1944, p. 1178.
20. Nichols Engineering Co., Bulletin No. 238 A, *Sludge Furnaces.*
21. Roberts, J. M., and Roddy, C. P., "Recovery and Reuse of Alum Sludge at Tampa," *Journal American Water Works Association,* 1960, p. 857.
22. Rose, J. L., "Removal of Phosphorous by Alum," FWPCA Seminar on Phosphate Removal, Chicago, Ill., June, 1968.
23. Sawyer, C. N., "Some New Aspects of Phosphates in Relation to Lake Fertilization," *Sewage and Industrial Wastes,* 1952, p. 768.
24. Schepman, B. A., and Cornell, C. F., "Fundamental Operating Variables in Sewage Sludge Filtration," *Sewage and Industrial Wastes,* 1956, p. 1443.
25. Schmid, L. A., and McKinney, R. E., "Phosphate Removal by a Lime-Biological

Treatment Scheme," *Journal Water Pollution Control Federation,* 1959, p. 1259.
26. Sebastian, F., and Sherwood, R. "Clean Water and Ultimate Disposal," *Water and Sewage Works,* Aug., 1969.
27. Sharman, L., "Polyelectrolyte Conditioning of Sludge," *Water & Wastes Engineering,* 4:8 (1967).
28. Slechta, A. F., and Culp, G. L., "Water Reclamation Studies at the South Tahoe PUD," *Journal Water Pollution Control Federation,* 1967, p. 787.
29. Smith, C. E., "Use and Reuse of Lime in Removing Phosphorus from Wastewater," Annual Meeting, National Lime Assoc., Phoenix, Arizona (Apr., 1969).
30. Tenney, M. W., and Cole, T. G., "The Use of Fly Ash in Conditioning Biological Sludges For Vacuum Filtration," *Journal Water Pollution Control Federation,* 1968, p. R281.
31. Tenney, M. W., *et al.,* "Chemical Conditioning of Biological Sludges for Vacuum Filtration," *Journal Water Pollution Control Federation,* 1970, p. R1.
32. Tomas, C. M., "The Use of Filter Presses for the Dewatering of Sewage and Waste Treatment Sludges." Presented at the Forty-second Annual Conference of the WPCF, Dallas, Texas (Oct. 1969).

10

Other Methods

ION EXCHANGE

A standard use of the ion exchange process is the removal of inorganic ions from water, as evidenced by the large number of water softening and deionization plants throughout the world. In applying the process to a tertiary sewage effluent, the goal is to remove inorganic nutrients. Many organic materials also may be removed by ion exchange and may be difficult to remove from some resins. This organic fouling of the resin hindered many of the early attempts to treat sewage effluents by ion exchange. The organic load on the ion exchange resin may be greatly reduced by preceding the column with activated carbon adsorption. Another factor which further reduces the organic fouling problem is the development of porous resins which have the ability to reversibly adsorb organic materials. In the early synthesis of ion exchangers, much emphasis was placed on obtaining the maximum number of exchange groups per unit of volume, and little attention was paid to the porosity of the resin. As a result, many of the early synthetic resins had high initial capacities which declined rapidly, because of fouling by ions which could not diffuse freely out of the interior of the resin.

Many of the reported experimental applications of ion exchange for sewage treatment have dealt primarily with treatment of secondary effluent. One exception is the study by Rand and Nemerow (1965) in which presettled raw sewage was passed through an anion exchange material (Resex 42). Excellent reductions of ABS, BOD, and phosphate were obtained, but physical plugging of the bed limited the length of run to 62 bed volumes. Only one run was reported, and no study of regeneration of the resin was made.

One of the first reported attempts to remove nitrogen from sewage by ion exchange was that by Gleason and Loonam (1933). The process consisted of four major steps:

1. Removal of suspended solids from raw sewage by coagulation with lime and iron compounds.
2. Sludge disposal by vacuum filtration and incineration.
3. Removal of nitrogen compounds by ion exchange with zeolite.
4. Regeneration of the zeolite with a salt solution and the subsequent recovery of ammonia from this solution.

The nitrogen, in the form of ammonia, was exchanged for sodium. The zeolite was regenerated with a salt solution containing 20 percent sodium chloride. Salt solution requirements were equal to 1 percent of the throughput. Approximately 16,000 lb of zeolite were required per million gallons of sewage treated. It was estimated that such a plant, with a 50-mgd capacity, could be installed at about two thirds the cost of an activated sludge plant. Unfortunately, no data were presented to show the initial exchange capacity of the resin and the exchange capacity after several regenerations.

Nesselson (1953) studied the use of ion exchange on secondary sewage effluent which had been passed through an anthracite coal filter. Strong base anion exchangers were found to satisfactorily remove nitrate by chloride anion exchange, but the handling of the large volume of waste regenerant (about 7 percent of the throughput volume) presented a difficult problem. He found the anion exchange capacity at a low flow rate, 1.2 gpm/ft^3, to be only 18 percent less than at a high flow rate 25 gpm/ft.^{3} Approximately 150 bed volumes of effluent were treated, on the average, prior to the effluent NO_3–N concentration increasing to 1 mg/l. Regeneration was accomplished with 40–70 lb of sodium chloride per cubic foot of anion exchange resin.

Eliassen and Wyckoff (1964) applied sand-filtered secondary effluent to anion exchange resin for nitrate and phosphate removal. Removals were good (75 percent nitrate and 97 percent phosphate removal) to over 200 bed volumes passed. Significant amounts of COD, ABS, and color were also removed. The efficiency of the resin was reduced by a high chloride and sulfate concentration in the sewage being treated. Organic fouling also occurred, as evidenced by a decreased throughput to phosphate breakthrough with continued use of the resin. Regeneration of the resin was accomplished with 2 bed volumes of 10 percent sodium chloride. It was estimated that the cost of salt for regeneration would be $40/mg million gallons of effluent treated. An attempt was made to apply unfiltered secondary effluent directly to the resin, but the bed plugged rapidly, and it was difficult to completely clean the resin by backwashing.

Eliassen, Wyckoff, and Tonkin (1965) attempted to apply the above laboratory scale ion exchange system on a larger pilot scale. Phosphate removal was only about 70 percent on the pilot scale, rather than the 97 percent obtained in the laboratory studies. However, nitrate removal was higher (90 percent) in the pilot-scale unit. The authors attributed the differences in the laboratory and pilot-scale units to a higher chloride concentration in the sewage and to mechanical problems encountered in preliminary runs with the pilot plant. They estimated the total cost as $207/mg exclusive of the cost of brine disposal.

Polio and Kunin (1968) investigated a modification of the Desal (Rohm and Haas Co.) process for treatment of secondary effluent. A weak-base anion exchange resin is converted to the bicarbonate form and the secondary effluent is treated by the resin to convert the inorganic salts. During this alkalization step, considerable adsorption of organics also occurs. After this step, the alkalized sewage is flocculated, aerated, and then lime-softened which precipitates more organic matter. The treated effluent can then be dealkalized by a weak cation resin to reduce the dissolved salt content and to remove ammonia. The anion resin is regenerated with aqueous ammonia and the cation resin with aqueous sulfuric acid. The resin did not appear to be fouled by the organics. The ammonia regeneration appeared to elute a great deal of the adsorbed organic matter during regeneration. The final effluent when treating secondary effluent had a COD of 17 mg/l, ammonium less than 5 mg/l as $CaCO_3$, and phosphates less than 1 mg/l as $CaCO_3$. The estimated capital and operating cost was about $180/mg exclusive of waste regenerant disposal.

Slechta and Culp (1967) investigated the use of a porous cationic resin (Duolite C25) for removal of ammonia nitrogen from tertiary effluent at the South Tahoe Water Reclamation Plant. About 400 bed volumes of carbon column effluent were passed through the ion exchange resin prior to breakthrough to 1 mg/l ammonia nitrogen at a service flow rate of 3 gpm/ft^3. The bed was regenerated with a 10 per cent sodium chloride solution. The cost of the salt to regenerate the resin was estimated at $45/mg. The disposal of the brine, the volume of which was estimated at 0.5 percent of the flow treated, made the process uneconomical in the Lake Tahoe basin.

The disposal of waste regenerants have been major problems associated with the use of standard ion exchange processes for wastewater treatment, as illustrated by the examples presented above. The cost of the nonselective ion removal provided by conventional resins is rarely justifiable from a pollution control standpoint. Thus, the recent development of ion exchange processes selective for ions of concern in control of pollution and which permit the recycling of the regenerants is of interest.

Selective Ion Exchange for Ammonia Removal

Battelle Northwest (1969) has developed a selective ion exchange process for ammonia nitrogen removal. The process employs a natural zeolite, clinoptilolite, which is selective for ammonium ions in the presence of sodium, magnesium, and calcium ions. Regeneration of the exhausted clinoptilolite is accomplished with a lime slurry which provides hydroxyl ions to react with the ammonium ions to yield an alkaline aqueous ammonia solution. This ammonia solution is processed through an air stripping tower to remove the ammonia, permitting recycling of the regenerant to the zeolite bed. Because the regenerant is not discarded, the process generates no liquid wastes. A mobile pilot plant has been operated at several locations in the United States and has generally provided 93–97 percent ammonia removal. Normally about 200 column volumes of effluent can be processed to full column loading. Regeneration can be achieved with about 8 column volumes of recycled regenerant applied at the rate of 10 column volumes per hour. Hydraulic loading rates of 6–8 gpm/ft^2 of bed area were used for normal operation. Preliminary indications are that the secondary effluent would require clarification by plain filtration prior to ion exchange to prevent fouling of the zeolite.

Battelle Northwest characterized the selectivity of the clinoptilolite and found that the zeolite is ammonium selective in systems containing calcium, sodium, and magnesium but prefers potassium to ammonium. Selectivity coefficients were determined at several points on an isotherm and are plotted against corresponding ratios of cation normalities in the equilibrium solution in Figures 10–1 and 10–2 for the specific Hector Clinoptilolite used. These curves are useful for

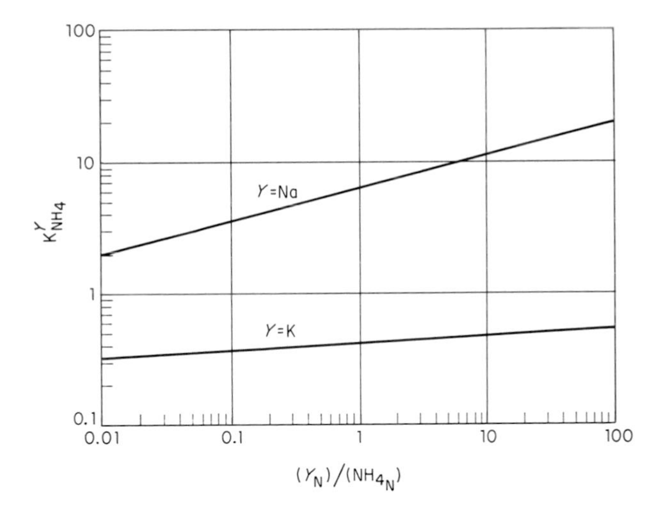

Figure 10–1 Selectivity coefficients vs. Concentration ratios of sodium or potassium and ammonium in the equilibrium solution with Hector clinoptilolite at 23°C for the reaction $(Y)_z + (NH_4)_N = (NH_4)_z + (Y)_N$. *(From Battelle Northwest, 1969)*

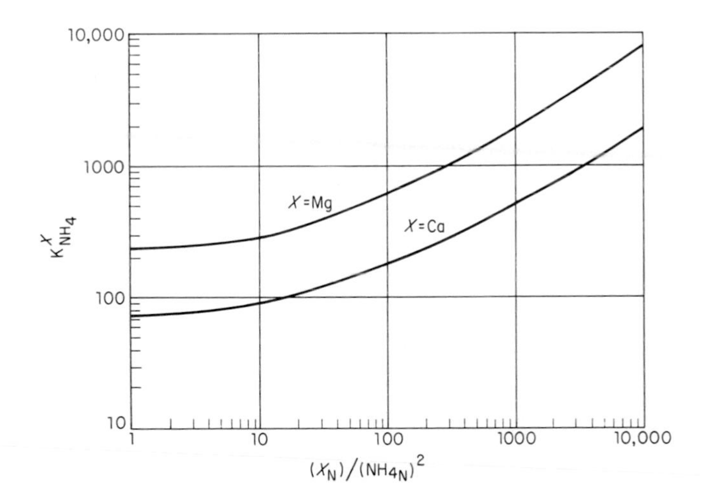

Figure 10–2 Selectivity coefficients vs. concentration ratios of calcium or mangesium and ammonium in the equilibrium solution with Hector clinoptilolite at 23°C for the reaction $(X)_z + 2(NH_4)_N = 2(NH_4)_z + (X)_N$. *(From Battelle Northwest, 1969)*

estimating the ammonium ion loadings for different waste compositions.

The regeneration of the zeolite was improved by adding sodium chloride to the regenerant—0.1 *N* NaCl saturated with $Ca(OH)_2$. The elution of ammonia was accelerated by the sodium ions and the sodium sharpened the ammonia elution curve due to the more rapid exchange rate of sodium relative to calcium. Some calcium carbonate and magnesium hydroxide floc accumulates in the regenerant and is removed by sedimentation between cycles. Very preliminary cost estimates indicated a total process cost on the order of $80/mg, a cost significantly lower than demonstrated by the conventional ion exchange sysems evaluated by earlier investigators, especially considering that the cost of regenerant disposal was generally not included in the earlier studies. Although the cost of the Battelle process is about five times that of the ammonia stripping process discussed earlier, the degree of removal provided is higher.

The potential merits of the system have led to preparation of a preliminary design report for a 7.5 mgd system at the South Tahoe Water Reclamation Plant. This selective ion exchange plant for ammonia removal would supplement the existing ammonia stripping tower in warm weather and will supplant it in cold. Final trace amounts of ammonia remaining in the wastewater after ion exchange treatment would be removed by breakpoint chlorination. The proposed design criteria are given in Table 10–1.

There would be a total of twelve ion exchange beds, nine of which would be in service and three in regeneration at all times. Figure 10–3 is a typical section through an ion exchange bed. The design flow rate is 6 gpm/ft^2. Because the *p*H of the lime regenerant solution is 11+, calcium carbonate deposition may occur, and the inlet and outlet screens are removable for cleaning without disturbing the clinoptilolite. The beds are designed so that clinoptilolite can be added or removed in slurry form.

A transfer tank would be provided for washing and hydraulically transferring the zeolite to the beds.

At design flow the service cycle for a set of 3 beds would be about 29 hr, and a set of 3 beds would have to be regenerated about every 9 hr. Elution would take place in two phases. In the first phase, elutrient water from a previous regeneration with an ammonia content of about 100 mg/l would be recirculated through the beds until the ammonia in the solution reached a concentration of about 600 mg/l. Throughout the recirculation, makeup lime would be added to maintain the *p*H at 11, and salt would be added as necessary. In the

Table 10-1 Proposed Design Criteria for Ammonia Removal by Selective Ion Exchange and Breakpoint Chlorination.

Plant Flow Capacity	7.5 mgd
Ammonia Concentrations	
Influenct	3–20 mg/l NH_3–N
Ion exchange effluent	0.5–1.0 mg/l NH_3–N
Following breakpoint chlorination	0.0 mg/l NH_3–N
Ion Exchange Beds	
Length of service cycle	175 bed vol
Service cycle loading	6 bed volumes/hr
Clinoptilolite bed depth	8 ft
Bed diameter	12 ft
Exchange Bed Regeneration	
Quantity of elutrient	8 bed vol
Elution rate	10 bed volumes/hr
Elutrient solution	
Calcium oxide	500 mg/l
Sodium chloride	0.2 N
*p*H	11.5–12
Heated Ammonia Stripping Tower for Regenerating Elutrient Solution	
Capacity	300 gpm
Air/water ratio	300 cfm/gpm
Hydraulic loading	3.5 gpm/ft^2
Packing height	24 ft
Efficiency per pass	85%
Number of passes	2
Temperatures	
Elutrient solution	74°F
Stripping tower air	74°F
Breakpoint Chlorination Dosage	10 mg Cl_2/mg NH_3–N

second phase, elution water with an ammonia concentration of about 10 mg/l would be recirculated through the bed until the ammonia concentration reached about 100 mg/l. Then the small amount of ammonia remaining in the bed would be distributed uniformly throughout the bed by means of passing one bed volume of ammonia-free water upflow through the bed. The elutrient solution, which had an ammonia concentration of 600 mg/l, would be held for air stripping during the regeneration of the next set of beds. Three elutrient storage tanks each with a capacity of about 100,000 gal would be required to store the various elutrient solutions. Such tanks would be equipped for manual removal of settled sludge. At design capacity about 1½ tons of lime per day are required as makeup for bed elution. Feeder-slaker capacity required to handle lime demands would be 800 lb/hr. About 2½ tons/day of sodium chloride would be used,

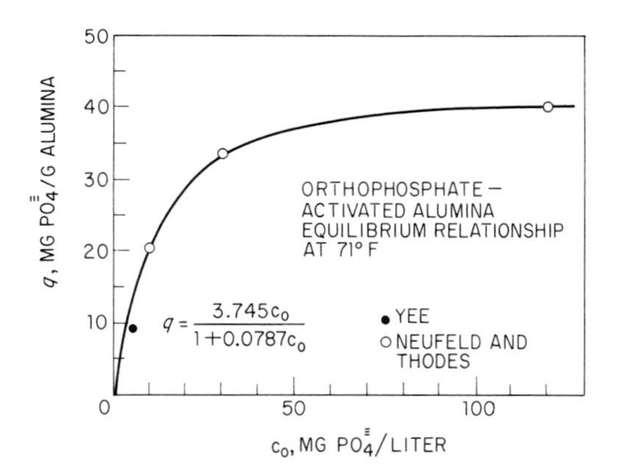

Figure 10–4 Equilibrium relationship for orthophosphate in solution and on solid alumina at 71°F. *(From Nuefeld and Thodos, 1969)*

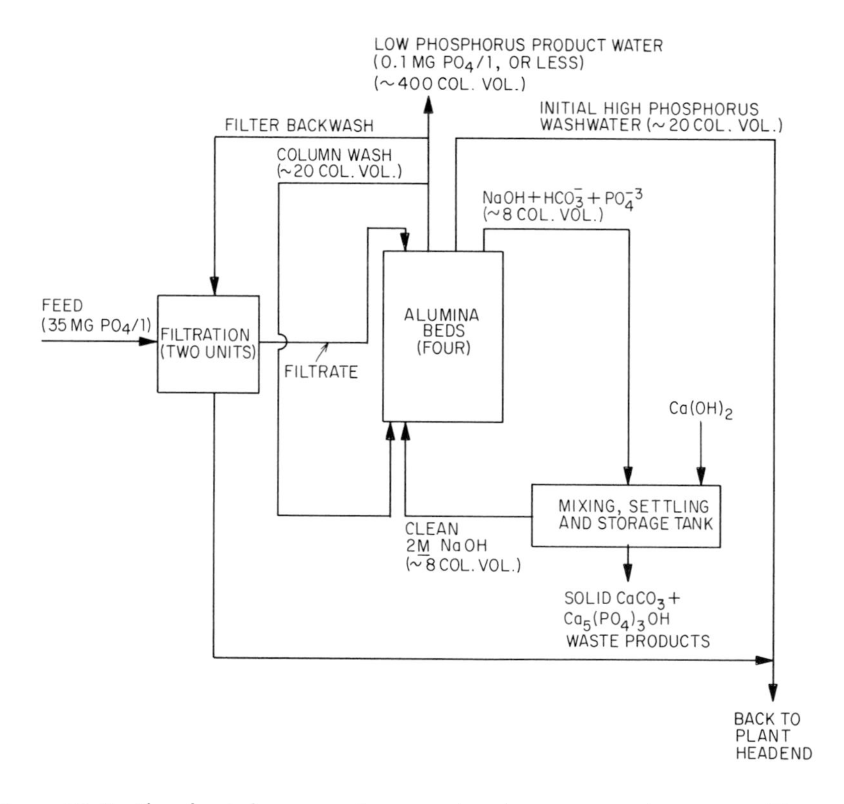

Figure 10–5 Flowsheet for a continuous phosphorus removal process. *(Courtesy Battelle Northwest)*

coagulation for some wastes on the basis of cost per million gallons of liquid treated while eliminating the voluminous chemical sludges often resulting from phosphate removal by chemical coagulation.

Ames (1969) has studied the feasibility of applying activated alumina to secondary effluents. He found that 400-column volumes of trickling filter plant effluent could be passed through 20–50 mesh, 275 m^2/g alumina column to reach 10 percent phosphorus breakthrough. If two columns were operated in series, more than 600-column volumes could be passed through the first before elution was required. Elution of the phosphorus was accomplished with 8-column volumes of 2 *M* NaOH followed by 20-column volumes of wastewater. The phosphorus can be removed from NaOH regenerant by the addition of 1.25–6.25 g/l of lime, allowing the reuse of the NaOH. Thus only a small volume of solid waste is produced by the alumina process. Alumina losses of about 0.1 percent per elution cycle were indicated. A portable pilot plant has been constructed and is now being tested to scale-up the promising laboratory work conducted by Ames. Nuefeld and Thodos (1969) have developed an equilibrium relationship for orthophosphate in solution and on solid alumina (28 × 48 mesh alumina was used in their tests) as shown in Figure 10–4. Ames presented a flow sheet of a proposed process for continuous phosphorus as shown in Figure 10–5.

CHLORINATION

Oxidation of Organics

Chlorination for disinfection of effluents is a well-accepted method in the wastewater field. Disinfection practice is not included in this discussion, which is limited to other potential uses of chlorination as an advanced waste treatment unit process.

Chlorination is a potential means for reducing the BOD of an advanced waste treatment plant effluent. For example, Lynam, *et al.* (1969) report that a tertiary effluent BOD of 9.5 mg/l was reduced 74 percent by a chlorine dosage of 1.25 mg/l and a residual of 0.35 mg/l. Chlorination may provide an effective temporary means of reducing the soluble BOD in absence of adsorption equipment.

Meiners, *et al.* (1968) found that ultraviolet radiation will very significantly increase the rate of chlorine oxidation of organic material in highly nitrified effluents from biological sewage treatment plants. Also, they found the extent of oxidation which can be achieved by requirements are 7–7.5 kwh per pound of ozone produced for ozonation and 2.5–4 kwh/lb for the air treatment equipment.

Hewes and Davison (1969) have found that the reaction rate of ozone with organics in flocculated sewage effluent is dependent on the rate of ozone decomposition, which is independent of the subsequent reaction with organics. They found that decomposition of ozone in water is second order at *p*H values of 2–4, 3/2 to second order at *p*H 6, and first order at *p*H 8. The rate increases rapidly at *pH* values above 6. To evaluate the effects of ozonation of wastewater, they bubbled ozone through secondary effluent which had been subjected to chemical coagulation and filtration prior to ozonation. Figure 10–10 summarizes their results. For sewage at 10°C, the COD reduction appeared to be due to reactions with decomposition products of ozone which are probably short-lived free radicals and ions.

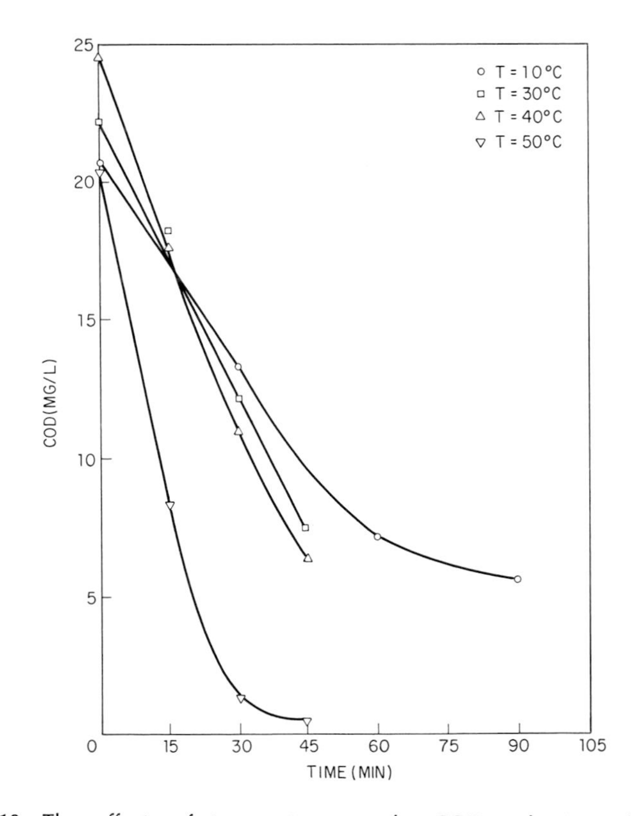

Figure 10–10 The effects of temperature on the COD reduction of pretreated secondary municipal wastewater effluents by ozonation. *(From Hewes and Davison, 1969)*

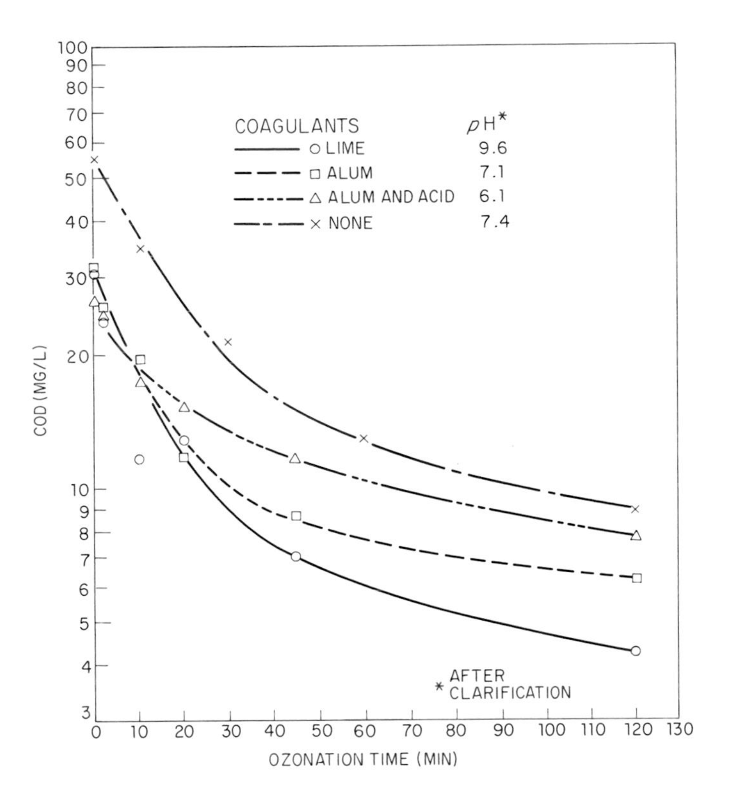

Figure 10–11 Effect of clarification on COD removal with excess ozone. *(From Huibers, et al., 1969)*

Huibers, *et al.* (1969) estimated the capital and operating cost of reducing a COD of 35 mg/l to 15 mg/l by ozonation to be $77/mg for a 10-mgd plant. They recommend chemical clarification of the secondary effluent prior to treatment with ozone for effluents having COD values above 40 mg/l. Figure 10–11 illustrates the effects of clarification as well as those of *p*H. The above cost estimate does not include the cost of the clarification step, which was estimated to add another $80/mg. They found that about an hour was required to reduce the COD from 35–40 mg/l to 15 mg/l while the half-life of ozone in water is about 20 min. As a result, increased utilization of ozone would be obtained by a six-stage cocurrent contacting system in which the ozone supplied per stage is equivalent to the amount that can be usefully consumed. Figure 10–12 illustrates the design concept by Huibers, *et al.* An operating *p*H of 6–7 was recommended as the optimum. Each stage of the contactor has a detention time of 10 min, or for bacterial growth. This quantity is an additional amount of about 30 percent over the stoichiometric amounts given in the above equations. From these considerations, the following formula may be used for estimating the total amount of methanol required:

$$C_m = 2.47\ N_0 + 1.53\ N_i + 0.87\ D_o$$

where C_m = required methanol concentration in mg/l
N_0 = initial nitrate-nitrogen concentration in mg/l
N_i = initial nitrite-nitrogen concentration in mg/l
D_o = initial dissolved oxygen concentration in mg/l

The value of C_m calculated above is somewhat conservative in that it does not make any allowance to the residual BOD entering the denitrification step. However, the portion of this residual BOD which can actually be utilized in the denitrification step is not known. Barth (1969) has suggested a methanol to nitrate-nitrogen ratio of 4.0 as a design guideline.

For an effluent containing 20 mg/l NO_3-N, 0.5 mg/l NO_2-N, and 2 mg/l D_o, the cost of adding the 52 mg/l of methanol indicated by the above equation is estimated at $16.30/mg based on a methanol cost of 3.8 cents/lb.

The early attempts to apply the denitrification concept involved passing the activated sludge mixed liquor into an anaerobic, mixed basin. However, it has been demonstrated by several recent investigators that the denitrification step can be combined with the columnar operations of granular activated carbon adsorption or granular media filtration with methanol fed to the column influent.

Parkhurst, *et al.* (1967) noted that denitrification was occurring in carbon columns being used for treatment of secondary effluents. It was found that nitrate reduction was limited to about 3 mg/l without the addition of supplemental organic feed. Methanol was added in amounts of 20 or 40 mg/l of COD per 25 mg/l nitrate-nitrogen with an increase of nitrogen removal to 15 mg/l noted. The total carbon column detention time was about 36 min, but similar results have been obtained at detention times as low as 5 min. Backwashing the column did not appear to interfere with the denitrification capacity of the column.

St. Amant and McCarty (1969) studied several media for use in a columnar operation to remove nitrates from drinking water supplies, including sand, activated carbon, volcanic cinders, gravel, coal and plastic trickling filter media. Sand and activated carbon were efficient but required regular backwashing to prevent clogging. The most satisfactory design appeared to be the use of 1-in. diameter gravel in a column providing 1 hr detention with methanol injected into the column influent. About 90 percent total nitrogen removal was achieved over a 9-month period without the need to clean the filter. Seidel and Crites (1970) have applied the same columnar approach to wastewater with similar results.

The rate of denitrification is temperature dependent, as would be expected for a biological process. The above filter maintained 90 percent nitrogen removal at temperatures as low as 12° but a filter with a 0.5 hr detention time suffered a drop in efficiency to 70 percent nitrogen removal at 12°C. The loss of biological solids from such a column would require a downstream filter to minimize the effluent solids content. A denitrification column 10 ft square and 6 ft high using 1-in. gravel has been built for tests of nitrate removal from irrigation return waters near Firebaugh, California.

Another approach to apply the principles of nitrification-denitrification has been developed by Barth (1969). Three separate stages of reaction basins and associated clarification and sludge return are used. A high-rate (approximately 2 hr detention) activated sludge process converts the carbonaceous material in primary effluent ot bacterial cells and CO_2 in the first stage. The solids generated in the first stage are removed by settling and recycled. The clarified effluent flows to a second stage which is a nitrification step. The nitrifying organisms are recycled to the second stage, permitting nitrification to be completed in about 3 hr detention. The third stage is denitrification of the clarified second-stage effluent, again with its own solids recycling system. Methanol is added to provide the proper ratio of carbon to nitrogen for denitrification. Laboratory results indicate 85–90 percent removal of nitrogen is possible with such a system. The practicality of consistently maintaining the necessary biological reactions and the related economics must be demonstrated on a plant scale before the potential of the process may be accurately evaluated.

POWDERED ACTIVATED CARBON

Although granular activated carbon has been applied to wastewater treatment on plant scale, powdered activated carbon has yet to be used other than in pilot plant studies. Powdered activated carbon has excellent adsorptive capacities and has been used widely in water treatment plants for removal of organic materials causing taste and odor problems. The lack of an efficient regeneration system is a major obstacle to widespread use in wastewater treatment due to the carbon doses required being much higher than in water treatment.

Davies and Kaplan (1965) studied a pilot-scale adsorption system for sewage effluents using powdered carbon. Powdered carbon doses of 240–600 mg/l were used. The powdered carbon was mixed with the effluent, and the carbon was then flocculated with the aid of a polyelectrolyte and separated from the effluent in a sludge-blanket settling system. About 90 percent of the equilibrium capacity of the carbon was reached in a 5-min mixing period. The carbon-treated effluent COD was about 5 mg/l. The spent powdered carbon was regenerated in a steam atmosphere at 750°F, with a 1-hr detention in the furnace. The estimated cost of the powdered carbon treatment was $67–79 per million gallons for a 10 mgd plant.

Beebe and Stevens (1967) applied powdered activated carbon to chemically coagulated and settled activated sludge effluent. The powdered carbon and coagulants (15 mg/l ferric sulfate and 0.5 mg/l polymer) were removed in a solids-contact clarifier. The clarifier effluent was passed through a sand filter for further clarification. Carbon doses of 140–180 mg/l were found to reduce the COD from 36–43 mg/l to 14–23.5 mg/l. They estimated that a powdered carbon dose of 385 mg/l would have been required to achieve the same results without the benefits of the solids-contact feature of the clarifier used. The cost of the chemicals used in the carbon step, based on a single use, were estimated at $119/mg. Regeneration of the carbon was not attempted.

A two-stage countercurrent process similar to that shown in Figure 10–16C was evaluated for application to secondary effluent by the FWPCA at Lebanon, Ohio. Cohen (1969) has reported that influent TOC concentrations of 15–25 mg/l were reduced to 1.5–3 mg/l. Based on a carbon dose of 200 mg/l and a carbon cost of 9 cents/lb, the cost of the carbon alone with no regeneration would be $150/mg. With efficient regeneration, Cohen estimated the total capital and operating cost of the powdered carbon process could be reduced to $109/mg.

The examples cited above certainly illustrate the need for an efficient regeneration technique if the use of powdered carbon is to be economically attractive. Several investigators are attempting to develop a regeneration system. Bloom, *et al.* (1969) reported on a transport-type system developed by FMC Corporation which they felt could regenerate powdered carbon at a cost as low as 1.5 cents/lb. The FMC technique involved injecting a carbon slurry (about 10 percent carbon) into a vertical gas-fired reactor at 1600°F with a small

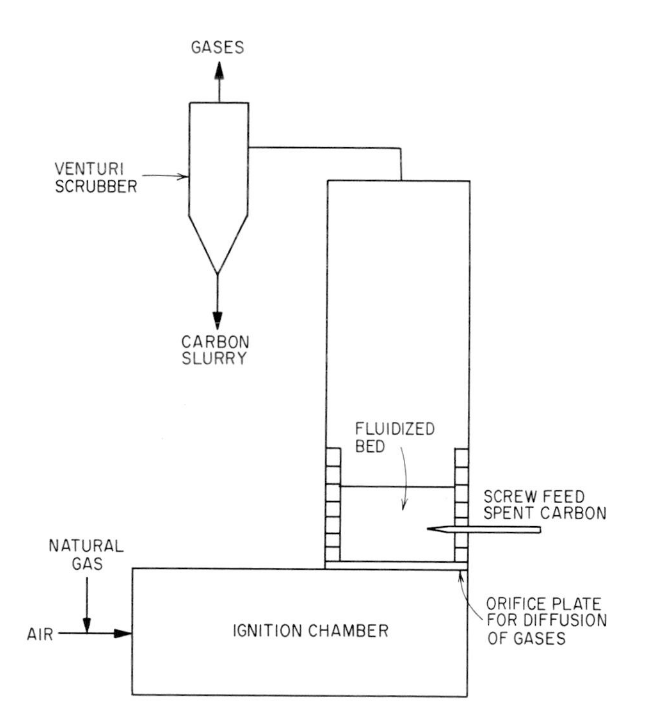

Figure 10–15 Simplified schematic of fluidized bed reactor for powdered carbon regeneration being developed by Battelle Memorial Institute.

airstream near the bottom of the reactor. The powdered carbon is efficiently regenerated with only a few seconds contact with the high temperatures. The gases are burned near the bottom of the reactor with the gases and the carbon leaving the reactor at the top and being recovered in a Venturi scrubber. The product slurry from the scrubber is recycled after being cooled.

A process showing promise in preliminary field tests is the fluidized bed approach (Figure 10–15) being developed by Battelle Memorial Institute. Preliminary tests with a 2-lb/hr unit indicated 85 percent recovery of the spent powdered carbon with 90–100 percent restoration of the adsorptive capacity. The feed carbon is thickened by centrifuging and contains about 25 percent solids when fed into the furnace. The carbon is fed by a screw conveyor into a fluidized sand bed. The gases from the ignition chamber are distributed by an orifice plate beneath the fluidized bed and carry the carbon to a Venturi scrubber or dry cyclone for recovery. Field test of units of 500-lb/day capacity, which were being initiated at the time of this writing, will better indicate the potential of this system and enable an accurate economic analysis.

Even should an efficient regeneration device become available, it remains to be seen whether or not powdered carbon systems will have a significant economic advantage over granular carbon systems.

COMBINED BIOLOGICAL-CHEMICAL TREATMENT

This section does not deal with a unit process but rather a combination of unit processes. Several investigators have shown that biological processes are capable of removing all soluble BOD found in raw sewage in reaction basins with very short residence times, on the order of 60 min or less. Unfortunately, the biological flocculation necessary to produce a settleable floc cannot be achieved within the same time period. However, chemical coagulation can readily convert the solids to a settleable state. It is the combination of high-rate biological treatment for removal of soluble BOD followed by chemical coagulation settling, and filtration for removal of solids and phosphate, that appears to be the integrated biological-chemical process offering the most potential, although others will be briefly discussed.

Tenney and Stumm (1965) made a basic study of complementary biological and chemical treatment in developing a flow sheet for a high-rate system. The process called for the raw waste to pass through a short-term aeration tank for removal of soluble BOD with the aeration tank effluent being chemically coagulated and settled. They felt, and logic strongly indicates, that the cost of the chemical treatment would be compensated for by a reduction in the cost of the biological treatment facilities resulting from shortened aeration times. The longer aeration times required for biological coagulation and flocculation are replaced by the chemical addition. They concluded that chemical removal of phosphate and chemical flocculation of the microorganisms present in the high-rate aeration tank effluent can be accomplished simultaneously. An important finding was that the optimum point for chemical flocculation of the microorganisms occurred at the same point in the microbial growth cycle which offered the maximum utilization of the biological capabilities of the organisms.

Eberhardt and Nesbitt (1968) reported on some continuous-flow lab-scale studies of the basic process Tenney and Stumm proposed. They found that essentially complete removals of soluble phosphorus are attainable by a combined biological-chemical process. They found that the alum dose required to produce satisfactory flocculation and clarification exceeded that necessary to produce a low, soluble phosphate residual. The alum addition did not adversely affect the performance of the activated sludge process with 2–2.5 hr detention in the aeration tank. The volume of sludge generated was approximately the same as generated by a typical conventional activated sludge process but the weight of solids was about doubled by use of the biological-chemical process. The sludge was reported to "dewater easily" although no quantitative data were presented on this point. Effluent filtration was found necessary to provide minimum phosphate residuals. Barth and Ettinger (1967) found the addition of alum or sodium aluminate to the aeration tank (6 hr aeration) of an activated sludge pilot plant to provide 74–98 percent phosphate removal. Davis and Love (1969) found a biological (1.6 hr aeration)—chemical process to be well suited for the treatment of a relatively weak municipal waste. Adding only 20 mg/l of alum and 0.5 mg/l polymer directly to the aeration tank provided a final effluent BOD consistently less than 24 mg/l. They found that an alum dosage considerably lower than that required for phosphate removal would provide good coagulation and flocculation of the microbial growth in the aeration tank.

Although not an intimate mixture of biological and chemical processes, the addition of lime to the primary clarifier influent offers some similar advantages in that it reduces the BOD loading on the aeration tank in an activated sludge plant, permitting a decrease in aeration tank size and cost. Albertson and Sherwood (1969) and Schmidt and McKinney (1969) have reported on studies of such a process. Enough lime is added to the primary to raise the *p*H to 9.5–10.0. This lime dose is usually adequate to remove about 80 percent of the phosphates from the raw sewage. The carbon dioxide liberated in the activated sludge aeration tank depresses the *p*H to 8.0–9.0. In some instances, the resulting sludge in the primary basin has proven to be very difficult to dewater.

Humenick and Kaufman (1970) have studied a system using a biological reactor with a volumetric BOD loading of 400–500 lb per 1,000 cu ft of aeration basin capacity (aeration times of 20–60 min) followed by settling and lime or alum coagulation and settling. Recycling of chemical solids to the aeration tank was found to control the settleability of the activated sludge and provided SVI values of less than 100, even with the very high organic loading used. The process provided over 90 percent BOD reduction and 95 percent phosphate removal without effluent filtration.

In many cases, use of very short-residence time activated sludge systems followed by chemical coagulation, settling, and filtration will produce a higher effluent quality for equal or lower costs than will the purely physical-chemical approach of coagulation, settling, filtration, and carbon adsorption. Such a system would occupy no more space than the physical-chemical, offsetting one of the major advantages of the physical-chemical approach. Much of the past disenchantment with the biological processes stems from periods of poor settling characteristics of the biological solids and gross carryover of solids in the plant effluent. Even during these periods, barring the presence of toxic materials, the soluble BOD is efficiently removed. Thus, the combined use of a short-residence time activated sludge system followed by chemical coagulation takes advantages of the best properties offered by each unit process. Development of this combination beyond the pilot plant stage will determine if full realization of this potential is practical.

PHYSICAL-CHEMICAL TREATMENT OF RAW SEWAGE

Physical-chemical treatment of raw sewage is, of course, not a unit process in itself but is rather an application of the unit processes described in this book to raw sewage or primary effluent, excluding the need for biological treatment. The reader is referred to two extensive literature reviews (Pearse, *et al.*, 1935, and Culp, 1967) of the subject rather than repeating a detailed history of the extensive work in this field. The concept is far from new, with published attempts appearing as early as 1740. The interest in the concept reached a peak in the 1880–1890 period in England with almost 200 plants employing chemical coagulation of raw sewage installed by 1890. Efficient chemical coagulation and settling of raw sewage may remove essentially all of the suspended solids but only 50–60 percent of the incoming BOD is related to these solids and is removed for most municipal wastewaters. As a result, mere chemical coagulation and settling may produce a very clear effluent but one that still has a very high BOD. Because of this fact and because of the large sludge quantities generated, most of the English plants that had been using chemical treatment were converted to biological types by 1910. The concept enjoyed a rebirth of interest about 1930 in the United States but faded from favor by the mid-1930's due to the same problems experienced earlier in England and in light of the more economic removal of organics provided by biological treatment processes.

The concept of physical-chemical treatment has recently experienced another surge in interest in the United States. However, it appears that the increasing number of instances requiring phosphate removal and high degrees of organic removal may have now established an economical use for the concept. As Culp (1967) pointed out, where phosphate removal and high organic removals are required for a new plant, nonbiological treatment (consisting of chemical coagulation, sedimentation, filtration, and granular activated carbon adsorption) would cost about the same ($192/mg) as biological treatment followed by chemical treatment (coagulation, settling, and filtration). The other cases where the economics are favorable include cases where sewage flow is intermittent or varies greatly, where space available for the plant is a limiting factor, or where industrial wastes that interfere with biological treatment are present. The recent studies on the subject by other investigators confirm these earlier conclusions.

Weber, *et al.* (1970) have estimated the cost of chemical coagulation, sedimentation, granular activated carbon adsorption and filtration of raw sewage at $160/mg. They reported that an effluent BOD of about 5 mg/l could be achieved as could 90 percent removal of phosphate in the system. The final effluent turbidities of about 1 JU achieved by Weber, *et al.*, indicate essentially complete removal of suspended solids. Biological activity in the carbon columns in physical-chemical systems applied to raw sewage is much higher than when applied to biologically treated effluents, due to the higher soluble BOD applied to the carbon. However, these biological growths appear to enhance the overall capacity for removal of organics by stabilizing a portion of the incoming organics. Weber, *et al.*, reported very high capacities of 60 percent TOC removal by weight with significant capacity yet remaining. This benefit is not without an associated problem in that anaerobic conditions in the carbon columns give rise to hydrogen sulfide in the plant effluent. This problem has been noted in every investigation known to the authors. Weber, *et al.*, reported that injection of oxygen in the carbon column influent provided satisfactory control of the problem. Addition of hypochlorite to the carbon column influent was only moderately effective and aeration of the influent tended to produce a biological floc which rapidly plugged and fouled the adsorbers. The biological growths also led to frequent sloughing of biological solids from the carbon columns, leading Weber to suggest that the filtration step follow the carbon adsorption step (Figure 10–16*A*). Frequent backflushing of the carbon columns has been suggested by others as a means of controlling the hydrogen sulfide generation.

Rizzo and Schade (1969) reported on a design in which phosphate

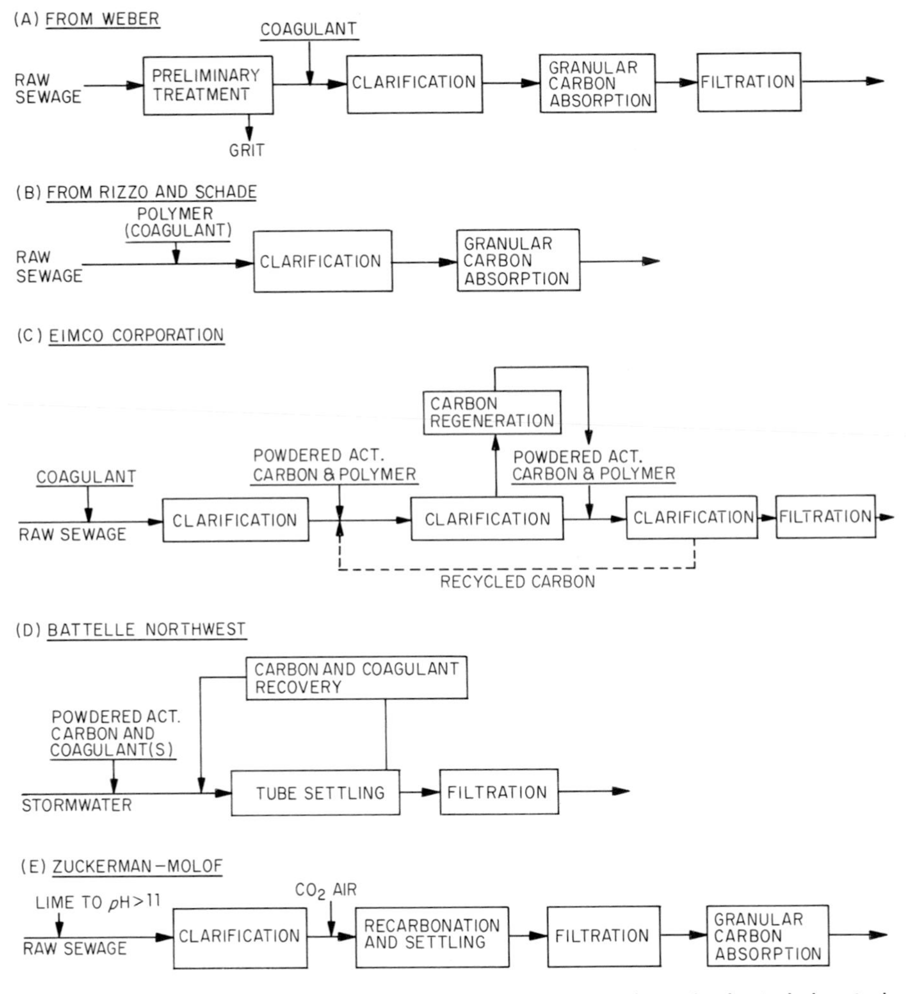

Figure 10–16 Simplified schematics as illustrative examples of physical-chemical treatment systems under development for application to raw sewage.

removal was not required initially. The raw sewage was coagulated with an anionic polymer, clarified by settling, and then applied to columns of granular activated carbon columns (Figure 10-16*B*). An average BOD of 8 mg/l and suspended solids of 7 mg/l were achieved in pilot tests with a carbon contact time of 33 min. Reduction of the contact time to 23.4 min increased the average BOD to 11 mg/l and a decrease to 14 min increased the average BOD to 21 mg/l. The carbon dosage for the plant design selected was estimated at 500 lb/mg. A 10-mgd plant has been designed for installation at Rocky River, Ohio. Hydrogen sulfide generation in the carbon columns was noted in the pilot studies but frequent backflushing of the columns reduced the problem.

Zuckerman and Molof (1970) of New York University investigated a process (Figure 10-16*E*) providing lime coagulation of raw sewage, sedimentation, recarbonation, filtration, and granular activated carbon adsorption. They claimed that the lime coagulation created a high *p*H, which hydrolyzed the larger soluble organic material to more adsorbable smaller molecules. Weber (1970) has taken exception to the hydrolysis theory.

Some work is under way to evaluate the feasibility of using powdered activated carbon in physical-chemical treatment schemes. The Eimco Corporation is examining the system shown in Figure 10-16C on a 100,000 gpd pilot scale in Salt Lake City, Utah. Preliminary tests of the same system have been made by the FWPCA in tests at Lebanon, Ohio, and these tests indicate that effluent TOC values of 4–10 mg/l may be possible. A key factor in the economics of such a system is the ability to efficiently recover and reuse the powdered activated carbon, which has yet to be demonstrated on a plant scale, as noted earlier. Battelle Northwest is investigating the use of powdered activated carbon for treatment of storm overflow in the system shown in Figure 10-16*D*. In this case, the coagulant (alum) and the powdered activated carbon are added with only a single clarification step following chemical treatment. The powdered carbon and alum are both recovered and recycled. The treatment system has a very low overall residence time due to the use of the tube settling concept and, should pilot tests be successful, it may be adaptable to raw municipal wastes as well as to storm water overflows.

There is little doubt that that physical-chemical treatment of raw sewage will play an important part in waste treatment in the future due to the increasing instances of required phosphate removal and BOD effluent concentrations lower than achievable with secondary processes, the associated need for high degrees of reliability, and the need to minimize space requirements for waste treatment facilities in increasingly complex urban environments. A physical-chemical plant using a process as outlined by Weber occupies only about 25 percent of the land area required by a biological treatment plant of the same capacity.

Weber, *et al.*, have estimated the operating and capital costs of a physical-chemical plant of the type shown in Figure 10-16*A* to be \$160/mg for a 10-mgd plant. Although this is about \$50–60/mg more than for a conventional secondary plant it is about $90–100 less than the cost of conventional secondary treatment followed by an advanced wastewater treatment plant providing chemical coagulation, sedimentation, filtration, and granular carbon adsorption. These economics are apparently based upon obtaining a readily dewaterable sludge from the combined step of chemical sludge and primary sludge removal. As noted in Chapter 9, the cost of dewatering such a combined sludge may be more than $60/ton in some cases. Under these circumstances, the overall costs could increase to about $300/mg. This example clearly illustrates the impact that sludge dewatering can have on the relative economics of various liquid-treating schemes.

ELECTROLYTIC TREATMENT

A great deal of interest was expressed prior to 1930 in electrolytic treatment of wastewaters. Several municipal plants were built, but all were eventually abandoned due to high costs and questionable efficiency. Recently, interest has been revived by work in Norway in which seawater is mixed with wastewater to increase conductivity and is passed between two electrodes before discharge to the ocean. The recent development of very stable, nonconsumable anodes has also added to the interest.

Electrolytic treatment is achieved in a cell composed of two metal strips dipped into water. Direct current is applied from a rectifier with the positive lead connected to the metal strip serving as the anode and the negative lead to the cathode. The electrical potential developed depends on the voltage applied by the rectifier. The higher the voltage, the greater the driving force to push electrons across the gap between the electrodes. If pure water is in the gap, very few electrons cross between electrodes. However, impurities increase the electrical conductivity, decreasing the required voltage. Chemical reactions occur at both electrodes. The cathode reactions result primarily in the decomposition of water to produce hydrogen gas which is given off in fine bubbles. The anode reactions are of more importance because oxidation occurs at the anode by four major means: (1) oxidation of chloride to chlorine or hypochlorite, (2) formation of highly oxidative materials such as ozone and peroxides, (3) direct oxidation by the anode, and (4) electrolysis of water to give gaseous oxygen. The fourth item is relatively ineffective, however, in oxidation of organics in wastewater. Most municipal wastewaters contain enough chloride ion that enough hypochlorite is produced to provide effective bacterial removal. The anode surface is electron-starved and will remove electrons, or oxidize, anything coming in contact with it. Thus oxidation occurs not only from the chemical oxidants generated but also at the surface of the anode.

The anode materials may be nonconsumable (noble anodes) or consumable (eroding anodes). Eroding anodes have been used in the past and may be constructed of aluminum or iron compounds. As they erode, the metal ion is released to serve as a coagulant. Due to the high operating cost of eroding anodes, the noble anodes appear to have more potential in wastewater treatment. Lead dioxide anodes do not erode when made anodic in wastewaters, appear to be the most effective anode for wastewater, and are commercially available.

Miller and Knipe (1965) concluded that the cost of electrolytic treatment was too high for removal of large percentages of secondary effluent COD. They estimated the cost of applying 18,000 c/l (coulombs/liter) as $900–2,400/mg, depending on the conductivity of the wastewater. This dosage provided 80 percent COD removal. Dosages of 1,200 c/l reduced coliform counts from 1,280,000/ml to 3,600/ml. Power requirements were about 5,000 c/l for 50 percent COD reduction. Reduction of power to this level would lower the overall costs by only about 20 percent according to the data of Miller and Knipe.

Although the above results indicate that electrolytic treatment is not competitive with activated carbon adsorption of 50-percent removal of COD from secondary effluents, recent tests at the Clark County (Nevada) Sanitation District indicate potential for economical reduction of secondary effluent bacteria and BOD. A noble lead dioxide anode manufactured by Pacific Engineering and Production Company of Henderson, Nevada, was used in these tests. Applied power of only 190 c/l reduced the coliform count from more than 10,000/ml to less than 100/ml while about 1,000 c/l was reported to reduce the BOD from 25–5 mg/l, although not significantly reducing the COD. The reported preferential removal of BOD over COD is shown in Figure 10-17. The substantial power required for 50-percent COD removal shown in Figure 10-17 confirms the conclusions of Miller and Knipe concerning COD removal. Unfortunately, Miller and Knipe did not report any BOD data. There is some question as to whether the BOD reductions shown in Figure 10-17 are real or are the result of bacterial inhibition in the BOD test due to presence of bactericides produced by electrolysis (chloramines, peroxides, etc.). The wastewater treated at Clark County contained 340 mg/l chloride as compared to 50 mg/l in the wastewater treated by Miller and Knipe. The cost of disinfection by electrolytic treatment at Clark

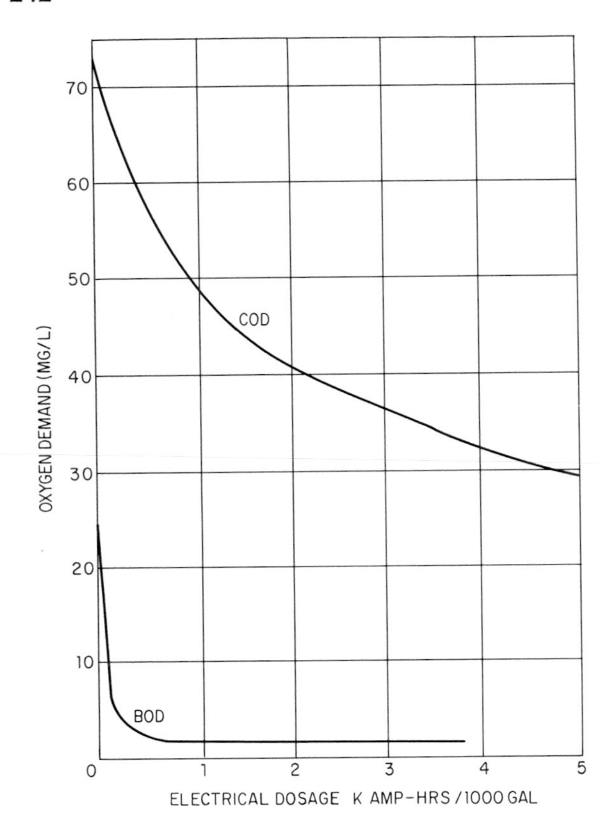

Figure 10–17 BOD and COD removal at Clark County, Nevada, by electrolytic treatment with lead dioxide anode. *(Courtesy Pacific Engineering and Production Co.)*

County has been estimated at $2/mg as compared to $3.60/mg for conventional chlorination. The incremental cost for BOD reduction from 25–5 mg/l has been estimated as $10/mg. These estimates were based on pilot tests and a power cost of 0.75 cents/kwh. Full-scale (12 mgd) facilities were being installed at the time of this writing. Evaluation of the full-scale operation will enable a more accurate evaluation of the potential of electrolytic treatment of wastewater which appears greatest for purposes other than gross COD removal.

References / ION EXCHANGE

1. Ames, L. L., "Evaluation of Operating Parameters of Alumina Columns for the Selective Removal of Phosphorus from Wastewaters and the Ultimate Disposal of Phosphorus as Calcium Phosphate," Final Report, FWPCA Contract No. 14–12–413 (1969).

2. Battelle Northwest, "Ammonia Removal from Agricultural Runoff and Secondary Effluents by Selected Ion Exchange," Robert A. Taft Water Reserach Center Report No. TWRC-5 (Mar., 1969).
3. Eliassen, R., and Wyckoff, B. M., "Progress Report, Reclamation of Reusable Water from Sewage," Technical Report No. 41, Dept. of Civil Engineering Stanford Univ. (August, 1964).
4. ———, and Tonkin, C. D., "Ion Exchange for Reclamation of Reusable Supplies," *Journal American Water Works Association,* 1965, p. 1113.
5. Gleason, G. H., and Loonam, A. C. "The Development of a Chemical Process for Treatment of Sewage," *Sewage Works Journal,* 1933, p. 61.
6. Kubli, H., "A Contribution to the Knowledge of Anion Separations by Means of Adsorption on Alumina," *Helv. Chem. Acta,* 1947, p. 453.
7. Mercer, B. W., *et al.,* "Ammonia Removal from Secondary Effluent by Selective Ion Exchange," *Journal Water Pollution Control Federation,* 1970, p. 95.
8. Nesselson, E. J., "Removal of Inorganic Nitrogen from Sewage Effluent," Unpublished PhD Thesis, Univ. of Wisconsin, Madison, Wisconsin (Dec., 1953).
9. Nuefeld, R. D., and Thodos, G., "Removal of Orthophosphates from Aqueous Solutions with Activated Alumina." *Environmental Science and Technology,* 1969, p. 661.
10. Polio, F. X., and Kunin, R., "Tertiary Treatment of Municipal Sewage Effluents," *Environmental Science and Technology,* Jan., 1968, p. 54.
11. Rand, M. C., and Nemerow, N. L., "Removal of Algal Nutrients from Domestic Wastewater, Part II. Laboratory Studies," Report prepared for New York Department of Health, Syracuse Univ., Dept. of Civil Engineering (Mar., 1965).
12. Slechta, A. F., and Culp, G. L., "Water Reclamation Studies at the South Tahoe Public Utility District," *Journal Water Pollution Control Federation,* 1967, p. 787.
13. Yee, W. C., "Selective Removal of Mixed Phosphates by Activated Alumina," *Journal American Water Works Assoc.,* 1966, p. 239.

CHLORINATION

1. Lynam, B., Ettelt, G., and McAloon, T., "Tertiary Treatment at Metro Chicago by Means of Rapid Sand Filtration and Microstrainers," *Journal Water Pollution Control Federation,* 1969, p. 247.
2. Meiners, A. F., Lawler, E. A., Whitehead, M. E., and Morrison, J. I., "An Investigation of Light-Catalyzed Chlorine Oxidation for Treatment of Wastewater," Robert A. Taft Water Research Center Report No. TWRC–3 (Dec., 1968).
3. Moore, E. W., "Fundamentals of Chlorination of Sewage & Wastes," *Water & Sewage Works Journal,* 1951, p. 130.
4. Smith, C. E., and Chapman, R. L., "Recovery of Coagulant, Nitrogen Removal, and Carbon Regeneration in Waste Water Reclamation," Final Report, FWPCA Demonstration Grant WPD–85 (June, 1967).

OZONATION

1. Diaper, E. W. J., "Microstraining and Ozonation of Sewage Effluents." Presented at the Water Pollution Control Federation Conference, Chicago, Illinois (Sept., 1968).
2. Hewes, C. G., and Davison, R. R., "The Kinetics of Ozone Decomposition and

Reaction with Organics in Water." Presented at the American Institute of Chemical Engineers National Meeting, Portland, Oregon (Aug., 1969).
3. Huibers, T. A., McNabney, R., and Halfon, A., "Ozone Treatment of Secondary Effluents from Waste Water Treatment Plants," Robert A. Taft Water Research Center Report No. TWRC–4 (Apr., 1969).

REVERSE OSMOSIS

1. Merten, U., Nusbaum, I., and Miele R., "Organic Removal by Reverse Osmosis." Presented at the American Chemical Society Symposium on Organic Residue Removal from Wastewaters, Atlantic City, New Jersey (Sept., 1968).
2. "Summary Report—Advanced Waste Treatment Research Program," Federal Water Pollution Control Administration Publication WP–20–AWTR–19 (1968).

NITRIFICATION—DENITRIFICATION

1. Barth, E. F., "Combined Treatment for Removal of Nitrogen and Phosphorus." Presented at FWPCA Technical Seminar on Nutrient Removal and Advanced Waste Treatment, Portland, Oregon (Feb., 1969).
2. Delwiche, C. C., "Biological Transformations of Nitrogen Compounds," *Industrial and Engineering Chemistry,* 1956, p. 421.
3. Eliassen, R., and Tchobanoglous, G., "Removal of Nitrogen and Phosphorus." Presented at the Twenty-third Purdue Industrial Waste Conference, Lafayette, Indiana (May, 1968).
4. Johnson, W. K., and Schroepfer, G. J., "Nitrogen Removal by Nitrification and Denitrification," *Jorunal Water Pollution Control Federation,* 1964, p. 1015.
5. Parkhurst, J. D., Dryden, F. D., McDermott, G. N., and English, J., "Pomona Activated Carbon Pilot Plant," *Journal Water Pollution Control Federation,* 1967, p. R70.
6. St. Amant, P. P., and McCarty, P. L., "Treatment of High Nitrate Waters," *Journal American Water Works Association,* 1969, p. 659.
7. Seidel, D. F., and Crites, R. W., "Evaluation of Anaerobic Denitrification Process," *Journal of the Sanitary Engineering Division,* 1970, p. 267.
8. Slechta, A. F., and Culp, G. L., "Water Reclamation Studies at the South Tahoe Public Utility District," *Journal Water Pollution Control Federation,* 1967, p. 787.
9. Wuhrmann, K., "Effect of Oxygen Tension on Biochemical Reactions in Sewage Purification Plants," Proceedings of the Third Conference on Biological Waste Treatment, Manhattan College, New York (Apr., 1960).
10. ———, "Nitrogen Removal in Sewage Treatment Processes," Fifteenth International Congress of Limnology Abstracts, University of Wisconsin, Aug., 1962.
11. "Objectives, Technology, and Results of Nitrogen and Phosphorus Removal Processes," *Advances in Water Quality Improvement,* University of Texas Press, Austin, Texas, 1968.

POWDERED ACTIVATED CARBON

1. Beebe, R. L., and Stevens, J. I., "Activated Carbon System for Wastewater Renovation," *Water & Wastes Engineering,* Jan., 1967, p. 43.

2. Bloom, R., Jr., Joseph, R. T., Friedman, L. D., and Hopkins, C. B., "New Technique Cuts Carbon Regeneration Costs," *Environmental Science and Technology,* Mar., 1969, p. 214.
3. Cohen, J. M., "Organic Residue Removal." Presented at FWPCA Technical Seminar on Nutrient Removal and Advanced Waste Treatment, Portland, Oregon (Feb., 1969).
4. Davies, D. S., and Kaplan, R. A., "Activated Carbon Treatment," *The American City,* Jan., 1965, p. 78.

COMBINED BIOLOGICAL-CHEMICAL TREATMENT

1. Albertson, C. E., and Sherwood, R. J., "Phosphate Extraction Process," *Journal Water Pollution Control Federation,* 1969, p. 1467.
2. Barth, E. F., and Ettinger, M. B., "Mineral Controlled Phosphorus Removal in the Activated Sludge Process," *Journal Water Pollution Control Federation,* 1967, p. 1362.
3. Davis, B., and Love, S., "Biological Chemical Combination Dramatically Cuts Treatment Costs," *Water and Pollution Control,* May, 1969, p. 35.
4. Eberhardt, W., and Nesbitt, J. B., "Chemical Precipitation of Phosphorus in a High-Rate Activated Sludge System," *Journal Water Pollution Control Federation,* 1968, p. 1239.
5. Humenick, M. J., and Kaufman, W. J., "An Integrated Biological-Chemical Process for Municipal Wastewater Treatment," unpublished paper (1970).
6. Schmid, L., and McKinney, R. E., "Phosphate Removal by a Lime Biological Treatment Scheme." Presented at the Water Pollution Control Federation Conference, Sept. 22–27, Chicago, Illinois (1968).
7. Tenney, M., and Stumm, W., "Chemical Flocculation of Micro-Organisms in Biological Waste Treatment," *Journal Water Pollution Control Federation,* 1965, p. 1370.

PHYSICAL-CHEMICAL TREATMENT OF RAW SEWAGE

1. Anonymous, "Sewage Treatment: Complete Process," *Chemical and Engineering News,* July 21, 1969, p. 8.
2. Culp, G. L., "Chemical Treatment of Raw Sewage," *Water & Wastes Engineering,* July, 1967, p. 61, and Oct., 1967, p. 54.
3. Pearse, *et al.,* "Chemical Treatment of Sewage," *Sewage Works Journal,* 1935, p. 997.
4. Rizzo, J. L., and Schade, R. E., "Secondary Treatment with Granular Activated Carbon," *Water and Sewage Works,* Aug., 1969, p. 307.
5. Weber, W., Hopkins, C. B., and Bloom, R., "Physiochemical Treatment of Wastewater," *Journal Water Pollution Control Federation,* 1970, p. 83.
6. Zuckerman, M. M., and Molof, A. H., "High Quality Reuse Water by Chemical—Physical Wastewater Treatment," *Journal Water Pollution Control Federation,* (1970), p. 437 (with discussion by W. Weber).

ELECTROLYTIC TREATMENT

1. Miller, H. C., and Knipe, W., "Electrochemical Treatment of Municipal Waste Water," U.S. Public Health Service Publication No. 999–WP–19 (1965).
2. Rhees, R. C., "The Pepcon Electrolytic Water Treatment Unit—What It Does," unpublished paper (1969).
3. Unpublished proposal to Clark County Sanitation District prepared by Pacific Engineering and Production Co., Henderson, Nevada (1970).

11

Laboratory Techniques

COLLECTION AND HANDLING OF SAMPLES

It is desirable to collect composite samples to reflect the performance of each unit process involved in the advanced wastewater treatment plant. Equipment is available to automatically collect these samples. Some parameters (turbidity, *p*H, TOC, phosphorus) may be measured and recorded continuously by automatic equipment. Grab samples will also be needed to evaluate, for example, the effects of flow variations on clarifier performance, the degree of carbon saturation, etc.

A great deal of attention should be given to the sampling techniques to be used. Careless sampling cannot be corrected by careful laboratory work. One of the frequent causes of error in sewage analysis is the failure to use a representative portion of the sample after it has been collected. Whenever the sample contains settleable material, it must be agitated gently to obtain a uniform distribution of solids. In many cases, solids may settle to the bottom of a sample container in the time period required to carry the sample from the sample point to the laboratory.

Of course, composite samples to be analyzed for organic content should be kept iced during the collection period and the analysis started as soon as the collection period is over.

TESTS FOR CONTROL OF PLANT OPERATION

The laboratory procedures for many of the tests involved to monitor the removal of organics, phosphorus, nitrogen, solids, etc. are found in the latest edition of *Standard Methods for the Examination of Water and Wastewaters* published jointly by the American Water Works Association, American Public Health Association, and the Water Pollution Control Federation, and no attempt will be made here to reproduce these procedures. A further discussion of analytical techniques may be found in *Chemistry for Sanitary Engineers* by Sawyer and McCarty (McGraw-Hill, New York). The discussion here will be limited to those tests which have not been routinely associated with the operation of water and waste treatment plants in the past or tests for which standard procedures are not readily available. The use of test results in the control of individual unit processes has been discussed in the related, previous portions of this book.

ACTIVATED CARBON ANALYSES

As noted in the earlier chapter on activated carbon adsorption, the apparent density test, the Iodine Number test, and adsorption isotherms play an important role in the control of carbon regeneration and in the design of carbon adsorption systems.

Apparent Density Test

1. Weigh a 100-ml graduate cylinder and record the weight in grams. A trip balance may be used. A vibrating shaker is attached to a ring stand above the graduate cylinder. The shaker may be obtained from the manufacturer of the activated carbon.
2. Pour a sufficient amount of the carbon to be tested into the funnel at the top of the shaker, place the graduate cylinder under the shaker, and fill to the 100-ml mark. The shaker should be adjusted so that the carbon fills the graduate cylinder at approximately 1 ml/sec.
3. Weigh the graduate cylinder and carbon and record the weight in grams.
4. To calculate the apparent density, determine the weight of carbon in the full graduate and divide by 100

   Weight carbon = Weight full graduate − weight empty graduate

$$\text{Apparent density} = \frac{\text{Weight Carbon}}{100} \ (\text{gm/ml})$$

The use of the apparent density test to control carbon regeneration is discussed in Chapter 8.

Iodine Number

The Iodine Number is defined as the milligrams of iodine adsorbed by one gram of carbon when the iodine concentration of the residual filtrate is 0.02 *N*.

Procedure:

1. Grind a representative sample of carbon until 90 percent or more will pass a 325-mesh sieve (by wet screen analysis).
2. Dry the sample for a minimum of 3 hr in an electric drying oven maintained at 150°C.
3. Weigh 1.000 g of dried pulverized carbon (see Note 2).
4. Transfer the weighed sample into a dry, glass-stoppered, 250-ml Erlenmeyer flask.
5. To the flask add 10 ml of 5%-wt HCl acid and swirl until carbon is wetted.
6. Place flask on hotplate, bring contents to boil and allow to boil for only 30 sec.
7. After allowing flask and contents to cool to room temperature, add 100 ml of standardized 0.1 *N* iodine solution to the flask.
8. Immediately stopper flask and shake contents vigorously for 6 min.
9. Filter by gravity immediately after the 6-min shaking period through an E & D folded filter paper.
10. Discard the first 20 or 30 ml of filtrate and collect the remainder in a clean beaker. Do not wash the residue on the filter paper.
11. Mix the filtrate in the beaker with a stirring rod and pipette 50 ml of the filtrate into a 250-ml Erlenmeyer flask.
12. Titrate the 50-ml sample with standardized 0.1 *N* sodium thiosulfate solution until the yellow color has almost disappeared.
13. Add about 2 ml of starch solution and continue titration until the blue indicator color just disappears.
14. Record the volume of sodium thiosulfate solution used.
15. Calculate the Iodine Number as follows:

$$\frac{X}{m} = \frac{A - (2.2B \times \text{ml of thiosulfate solution used})}{\text{weight of sample (grams)}}$$

$$C = \frac{N_2 \times \text{ml of thiosulfate solution used}}{50}$$

$$\text{Iodine Number} = \frac{X}{m} D$$

Where X/m = mg. iodine absorbed per gram of carbon
N_1 = normality of iodine solution
N_2 = normality of sodium thiosulfate solution
A = $N_1 \times 12693.0$
B = $N_2 \times 126.93$
C = residual filtrate normality
D = correction factor (obtained from Figure 11-1)

Notes on Method:

1. The capacity of a carbon for any adsorbate is dependent on the concentration of the adsorbate in the medium contacting the carbon. Thus, the concentration of the residual filtrate must be

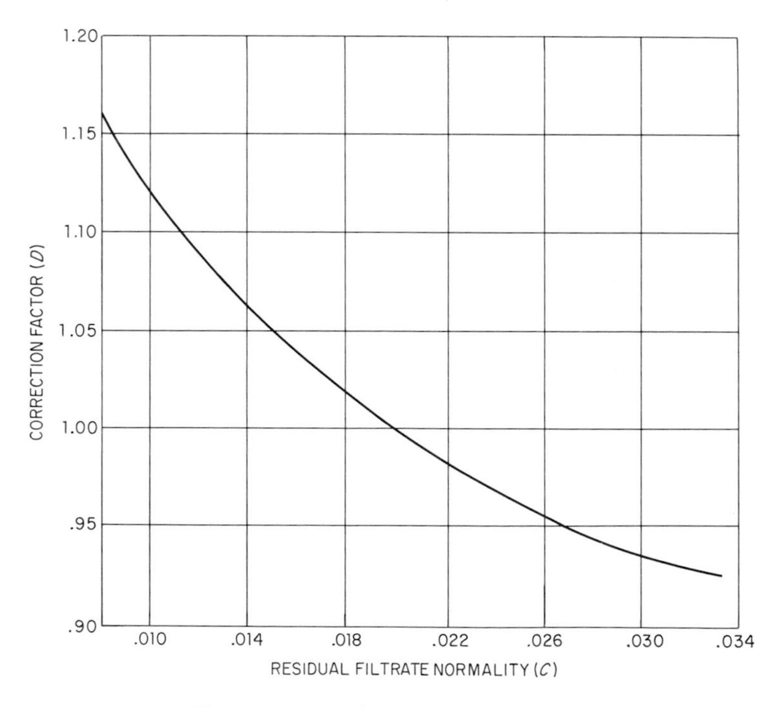

Figure 11–1 Iodine correction curve.

specified, or known, so that appropriate factors may be applied to correct the concentration to agree with the definition.

2. The amount of sample to be used in the determination is governed by the activity of the carbon. If the residual filtrate normality (C) is not within the range 0.008–0.035 *N*, given on the Iodine Correction Curve, the procedure should be repeated using a different size sample.
3. It is important to the accuracy of the test that the potassium iodide to iodine weight ratio is 1.5–1 in the standard iodine solution.

Reagents and Equipment:

Hydrochloric acid, 5%-wt. To 445 ml of distilled water add 55 ml of reagent-grade concentrated hydrochloric acid.

Sodium thiosulfate, 0.1 *N*. In a 1-l volumetric flask dissolve 24.82 g of reagent-grade sodium thiosulfate crystals ($Na_2S_2O_3.5\ H_2O$) in distilled water. Add about 0.1 g of reagent-grade sodium carbonate and dilute to the 1-l mark. This solution should be allowed to stand for a few days before standardizing.

Standardize with reagent-grade metallic copper. Dissolve about 0.2 g of copper, weighed to the nearest 0.1 mg, in 5 ml of concentrated nitric acid and boil gently to expel brown fumes. Dilute to about 20 ml with distilled water and add ammonia water dropwise until the solution is a deep blue color. Boil again until the odor of the ammonia is faint. Neutralize with acetic acid until the precipitate which forms with the acid dissolves and add 5 or 6 drops in excess. Again bring to boiling. Cool to room temperature. Add solid potassium iodide in sufficient amount to redissolve the copper iodide precipitate which forms. Titrate with sodium thiosulfate until the iodine fades to a light yellow color. Add starch indicator and continue the titration by adding the thiosulfate dropwise until a drop produces a colorless solution. Calculate the normality of the sodium thiosulfate as follows:

$$\text{Normality of sodium thiosulfate} = \frac{\text{weight copper}}{\text{ml thiosulfate} \times 0.06354}$$

Iodine solution. Dissolve 127 g of reagent-grade iodine and 191 g of potassium iodide in distilled water (see Note 3). Dilute to 1 l in a volumetric flask. To standardize the iodine solution, pipette 25.0 ml into a 250-ml Erlenmeyer flask and titrate with the standardized 0.1 *N* sodium thiosulfate. Use the starch indicator when the iodine fades to a light yellow color. Then finish the titration by adding the thiosulfate dropwise until a drop produces a colorless solution.

Calculate the normality of the iodine solution as follows:

$$\text{Normality of iodine solution} = \frac{\text{ml thiosulfate} \times \text{normality thiosulfate}}{25}$$

Starch indicator. Mix 1 g of soluble starch with a few milliliters of cold water. Pour the mixture into 1 l of boiling water and allow boiling to continue for a few minutes. This solution should be made up fresh daily for best results.

Filter paper. E & D, folded filter paper, 18.5 cm, No. 192.

Isotherms

Data for plotting isotherms are obtained by treating fixed volumes of the liquid to be tested with a series of known weights of carbon. The carbon-liquid mixture is agitated for a fixed time at constant temperature. After the carbon has been removed by filtration, the residual organic content of the solution is determined. From these measurements, all of the values necessary to plot an isotherm may be calculated.

The isotherm tests may be performed at room temperature unless the anticipated plant operation will be at a significantly different one. To determine the needed contact time for the isotherms, a preliminary experiment should be run in which fixed volumes of the wastewater are contacted with a fixed weight of carbon for 1-, 2-, 3-, and 4-hr periods. A contact time sufficiently long to insure a reasonable equilibrium should be chosen from these data for the isotherms.

The *p*H of the wastewater being tested will affect the carbon efficiency. Care should be taken to insure that the *p*H of the test sample is representative of anticipated plant-scale conditions.

In order to minimize the variable of carbon particle size in the isotherm test, the granular carbon should be pulverized so that 95 percent passes a 325 mesh screen. If suitable mechanical equipment, such as a ball mill, is not available to pulverize the carbon, a mortar and pestle can be used, although this is a tedious approach. The manufacturer of the carbon may also be able to supply pulverized material.

For the initial isotherm test, carbon dosages of 50, 100, 150, 200, and 300 mg/l of wastewater may be used. If the highest dosage is not adequate to effect the desired degree of treatment, higher dosages should be tried until that degree is achieved. If one of the intermediate dosages is sufficient to effect complete or satisfactory removal, one or two lower dosages should be tried. In order to obtain a satisfactory isotherm, as wide a range of organic removal should be obtained as is practical.

Isotherm Procedure:

1. Pulverize a representative sample of the granular carbon (a 10-20 g sample is usually adequate) so that 95 percent will pass through a 325-mesh screen. Oven-dry the pulverized sample for 3 hr at 150°C.
2. Obtain a representative sample of the wastewater to be tested. Suspended matter should be removed by filtration.
3. Transfer four different weights of the oven dried pulverized carbon to the test containers. Stoppered flasks or pressure bottles are satisfactory containers.
4. To one container, add 100 ml of wastewater from a delivery burette or graduate cylinder, and clamp the container on a mechanical shaker. The samples must be constantly agitated during the isotherm test and a mechanical shaker is desirable. A Burrell Wrist Action Shaker, Catalog No. 75-775, is satisfactory. Agitate the mixture for the chosen contact time. The bottles may be filled and placed on the shaker at 10- or 15-min intervals to give the analyst sufficient time to filter each sample immediately after the contact time has elapsed. The same volume of wastewater should be added to a container without carbon and subjected to the same procedure in order to obtain a blank reading.
5. After the chosen contact time has elapsed, filter the contents of the flask through either a laboratory pressure filter fitted with an asbestos disk or through a Buchner funnel containing a filter paper inserted in a filter flask connected to a vacuum. The blank should be filtered in the same manner as the other samples. It is desirable to discard the first and last portions of the filtrate and save only the middle portion for analysis.
6. Determine the organic content of the filtrate.
7. Tabulate the data as shown in Table 11-1. The residual solution COD concentration, c, is obtained directly from the filtrate analysis. The amount adsorbed on the carbon, x, is obtained by subtracting the value of c from that of c_0, the influent concentration. Dividing x by m, the weight of carbon used in the test, gives the amount adsorbed per unit weight of carbon.
8. On log paper plot c on the horizontal axis against x/m on the vertical axis and draw the best straight line through the points as illustrated in Figure 11-2.

Table 11-1 Tabulation of Isotherm Data.

m *Weight of Carbon (mg/100 ml) solution)*	c *Residual COD (mg/l)*	x *COD Adsorbed (mg)*	x/m *COD Adsorbed (per unit weight)*
0	40	—	—
5	27	1.3	0.26
10	19	2.1	0.21
15	14	2.6	0.17
20	10	3.0	0.15

From the isotherm, it is immediately apparent whether or not the desired degree of purification can be attained with the particular activated carbon tested. If a vertical line is erected from the point on the horizontal scale corresponding to the influent concentration (c_0) and the isotherm is extrapolated to intersect that line, the *x/m* value at the point of intersection can be read from the vertical scale. This value, $\left(\frac{x}{m}\right)C_0$, represents the amount of COD adsorbed per unit weight of carbon when that carbon is in equilibrium with the influent concentration. Since this should eventually be attained during column

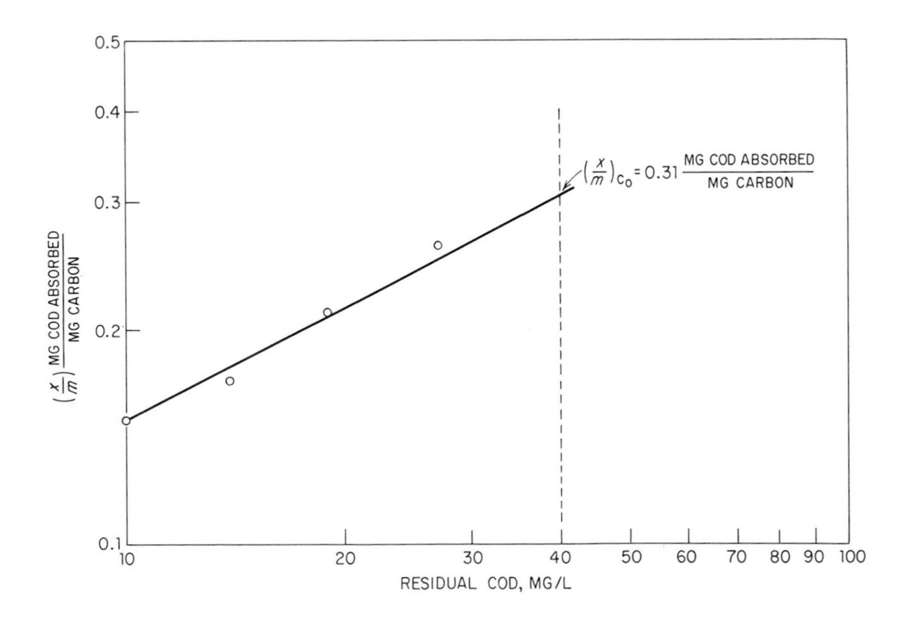

Figure 11–2 Illustrative isotherm.

treatment, it represents the ultimate capacity of the carbon. However, experience has shown that it is difficult to accurately predict the carbon dosage from isotherm tests for plant-scale carbon columns. Pilot carbon column tests conducted over several weeks are the only accurate means of determining the required carbon dosage.

The chief value of the isotherm test lies in comparison of various types of carbon and the ability to determine the quality of effluent achievable following carbon adsorption. The interpretation of isotherms was discussed in Chapter 7.

ASH ANALYSIS

Ash is defined as the mineral oxide constituents of the carbon. Heating a sample to a temperature of 1750°F in an oxidizing atmosphere will completely oxidize all carbon and convert the mineral constituents to their respective oxides.

Procedure:

1. Grind a representative sample of carbon until 90 percent or more will pass a 325-mesh screen (by wet screen analysis).
2. Weigh 1.000 g of pulverized sample into a weighed Vycor glass crucible without cover (see Note 1).
3. Place crucible and contents into a muffle furnace, set at 1750°F for 3 hr.
4. Remove from furnace, cool in desiccator, and weigh. (Save crucible and ash for iron analysis, if desired.)
5. Percentage of ash is calculated as follows:

$$\%\ \text{ash} = 100$$
$$(\text{final weight of crucible}) -$$
$$(\text{original weight of empty crucible})$$

Notes on Method:

1. Although it is preferable to use Vycor glass, 30-ml capacity, crucibles in this test, platinum or porcelain crucibles may also be used.

CALCIUM CONTENT OF RECALCINED LIME

The calcium content of recalcined lime may be determined as follows:

1. Place a small sample (about 1 g) of recalcined lime in a 103°C oven in a tared dish for 30 min. Cool in a desiccator.

2. Obtain weight of dried sample.
3. Dissolve sample in 1 l of distilled water.
4. To a 50-ml sample of the above solution, add 1 ml of 8 *N* potassium hydroxide.
5. Add one level teaspoonful of Calver II Indicator Powder (Hach Chemical Co. Catalog No. 281).
6. Titrate with standard Titraver solution (Hach Chemical Co. Catalog No. 205) to color change.
7. Calculate mg/l Ca: $\text{mg/l} = \dfrac{400 \text{ ml Titraver}}{\text{ml sample}}$
8. Calculate percentage of Ca: $\% \text{ Ca} = \dfrac{\text{mg/l Ca}}{\text{mg sample (100)}}$

Note: To prepare 8*N* potassium hydroxide, dissolve 448 g potassium hydroxide in 600 ml of distilled water and dilute to 1 l.

COAGULATION CONTROL BY JAR TESTS

The jar test is a laboratory procedure which affords a rapid means to determine the effects of chemical treatment on wastewater or sludge. Standard laboratory stirring equipment can be purchased from Phipps & Bird, Inc., or from Turbitrol Company. It consists of a series of six stirring paddles which can be rotated at a variable controlled

Figure 11–3 Jar testing. *(Courtesy Clair A. Hill & Assocs.)*

speed to mix contents of 1-l beakers as shown in Figure 11–3. This permits the simultaneous treatment and observation of six samples under identical mixing conditions in order to determine the relative merits of various chemical treatments, or to determine the optimum dosage of a particular chemical.

Jar tests can be used to determine the dosages of lime, alum, or other coagulant required for clarification of wastewater or the dosages needed to secure certain phosphorus removals. They can also be used as a screening test for polymers or other chemicals used as aids in the flocculation of wastewater or the dewatering of sludge. What constitutes the optimum dosage depends upon the objectives of treatment. In the case of clarification, size and density of the floc particles are important. Very often a small, dense floc, about pinhead size, will settle more rapidly and completely than a large, flaky floc, so that floc size is not in itself necessarily a good criterion to use. For phosphorus removal, a fine floc may give better results because of its greater surface area and the greater number of opportunities for contact afforded by the larger number of particles present between it and the phosphorus to be adsorbed.

Some simple tests of the water or sludge being treated, before or after mixing and settling of the samples are a valuable part of the procedure. Measurements of turbidity, color, *p*H, alkalinity, and other properties of the water or sludge are often useful or necessary to control or understanding of the tests.

Standard strength solutions of coagulant or sludge conditioner may be prepared by dissolving 1 g of the chemical in 100 ml of distilled water. A milliliter of this solution in the 1,000 ml of water being treated forms a dose of 10 mg/l. A dose of 1 mg/l is equal to 8.34 lb of the chemical per milion gallons of water being treated. One grain per gallon amounts to 17.1 mg/l.

In making jar tests it is desirable to duplicate insofar as possible actual full scale plant conditions. It is common to start the mixer at say 40–80 rpm, add the chemical and mix rapidly for about 30–60 sec, then reduce the paddle speed to 10–20 rpm and slow mix for an additional 4–30 min. At the end of the slow mix period, the paddles are then removed from the jars and the floc is allowed to settle. The time for all of the floc to settle to the bottom of the beakers is observed. If more than one chemical is added, the order of addition should be the same as in the plant. Chlorine often affects coagulation, flocculation, and settling of waters containing organics, and if prechlorination is practiced in the plant, this condition should be duplicated in the jar tests.

SLUDGE FILTERABILITY TEST

A Buchner funnel test provides the simplest means of determining the dewatering characteristics of a sludge and the effects of various types of chemical conditioning of the sludge. When evaluating chemical conditioning, the objective is primarily to determine a workable chemical dosage, not to optimize all aspects of the plant-scale dewatering system since there are many aspects which can be evaluated accurately only with the mechanical dewatering device involved. The first test described here is based upon using the volume of filtrate obtained in a uniform period of filtration as the measure of the filterability of the sludge. Other techniques are available, but the following procedure is simple and requires inexpensive, standard laboratory apparatus.

A 9-cm Buchner funnel is fitted with a rubber stopper to a 500-ml filtering flask. The flask is connected to a vacuum source in such a manner that the vacuum may be released rapidly at the end of the filtration time. The vacuum source should be capable of providing a reproducible vacuum in the range of 20–27 in. of mercury. The filter paper used should be a rapid filtering paper such as Whatman No. 1 or No. 4. The following procedure is recommended:

1. Start with a large, representative sample of the sludge and keep the sludge well mixed just prior to removing each portion for testing. Set up an analysis for the solids content of the sludge.
2. Seat filter paper in funnel with distilled water and vacuum. Disconnect vacuum and empty flask.
3. Pour 250 ml sludge onto seated filter paper, start vacuum, and observe time.
4. Release vacuum after 60 sec and remove filter.
5. Measure volume of filtrate and note appearance of filtrate.

If surface of sludge pulls down to a visible cake, repeat run using a 30-sec or shorter filtration time. If chemicals are added, calculate the chemical doses used as a percentage of dry solids. Plot the milliliters of filtrate as ordinates against chemical dose as abscissa. In some cases, other parameters such as *p*H may be of importance and should also be plotted. When the *p*H is the principal means of controlling the chemical dose, the volume of filtrate may be plotted as ordinate against the *p*H as abscissa.

A modification of the above test can be used to determine the specific resistance of the sludge to determine the pressure difference needed per unit of filtrate flow of unit viscosity per unit weight of filter cake. This modification has been used successfully to predict vacuum filter performance and provides more complete information than the simple Buchner funnel procedure described above. In this test, the volume of filtrate is recorded versus time of filtration. The specific resistance is calculated as follows:

$$r = \frac{26PA^2}{\mu c}$$

Where b = slope of $\frac{\text{time}}{\text{filtrate volume}}$ vs. filtrate volume (sec/cm^6)
P = filtration pressure (g/sq cm)
A = area, sq cm
μ = filtrate viscosity (poises)
c = weight of solids/unit volume of sludge prior to filtration (g/ml)
r = specific resistance, sec^2/g

Vacuum filter loading can then be predicted by the equation:

$$L = 35.7 \left(\frac{xcP}{\mu Rt}\right)^{1/2}$$

Where L = filter loading
R = $r \times 10^7$ (sec^2/g)
P = pressure, psi
c = solids deposited/unit volume of filtrate, g/ml
μ = filtrate viscosity (poises)
t = cycle time, min
x = fraction of cycle time for cake formation, min

References

1. Cox, C. R., *Laboratory Control of Water Purification,* Case-Sheppard-Mann Publ. Corp., New York, 1946.
2. *Standard Methods for the Examination of Water, Sewage, and Industrial Wastes,* 12th ed., American Public Health Assoc., 1790 Broadway, New York.
3. *Water Quality and Treatment,* American Water Works Association, 2 Park Ave., New York, 1950.

12
Operator Training

THE NEED FOR TRAINING

To meet the need for clean water, it is essential not only to build better and more advanced wastewater treatment facilities, but also to see that these plants are operated so as to realize their full potential benefits. The current tremendous interest in water quality offers to operators an unprecedented opportunity for public recognition of the importance of their work and of their accomplishments. The prevailing badly polluted conditions of the nation's lakes and rivers present many interesting challenges, and much personal satisfaction can be gained from the public service rendered in the correction of these problems.

The owner and the design engineer must not only finance and construct the physical plant, which is capable of meeting all water quality needs and regulations, but they must do their part in assuring proper plant operation and maintenance. Their role includes providing adequate funds for these purposes, and also in making all of the arrangements for the necessary operator training.

Operator training must strongly emphasize the necessity of operating wastewater treatment facilities so that the finished water continuously conforms to the established quality standards. This requires a full understanding of the physical plant and the plant processes and a working knowledge of hydraulics, bacteriology, and chemistry. It requires vigilant observation of the condition of plant equipment and development of a good program of preventive maintenance. In addition to the necessary technical knowledge, it requires determination on the part of the operating staff to maintain plant efficiency even in face of difficult operating conditions. Dedication and ingenuity are required.

Operators are also responsible for the appearance of the buildings and grounds, the keeping of good records, the dissemination of public information, and the cultivation of good public relations in their everyday contacts with people in the community.

There are presently only a small number of graduate engineers, chemists, and bacteriologists directly involved in plant operations. Most operators are hired locally and have little or no formal training directly related to their work. Plant operators must have access to textbooks or other reference materials on water and wastewater treatment. Every plant library must include a copy of *Standard Methods for the Analysis of Water, Sewage, and Industrial Wastes*. State and federal water pollution control agencies and many colleges and universities conduct short courses for the training of water and wastewater treatment plant operators, and federal grants are now available to assist operators who attend these schools. Operators must be encouraged and assisted in attending these meetings. Operators should be encouraged to join the local Water Pollution Control Association, attend its meetings, and read its publications. Trade journals should be made available to all plant personnel. Manuals on the operation of wastewater treatment plants are available from the Water Pollution Control Federation, *Public Works* magazine, *Water & Wastes Engineering* magazine, and other publishers.

ON THE JOB TRAINING

Despite all of these valuable training aids, the fact remains that on-the-job training is an absolute necessity for every operator in every plant. This training must start before the plant is placed in operation, and must continue as necessary to train new employees, and to keep experienced operators informed of new developments in the field or of changes and additions to the plant.

The value and importance of on-the-job training must be recognized and acknowledged by the owner, engineer, and operators. This means that the operators must be paid for the time spent in the training sessions and that outside instructors should be reimbursed for their work. Only in this way can this matter be given the attention and recognition which it deserves. The engineering consultant, the major equipment suppliers, and the construction contractor all have valuable background knowledge and information which must be conveyed to those persons who operate the plant. Their active support and participation in the training efforts can be assured by including provisions and payment therefore in their respective contracts with the owner. The engineer and the manufacturers are interested in seeing that the operators have the information needed to operate the equipment as they intended. The best and most efficient way to accomplish this is to conduct a formal school. Typically a school might consist of two 2-hr sessions per day for about 3 months in order to adequately cover a new plant operation. The two separate sessions, one in the morning and the other in the afternoon would make it possible for operators from all shifts to attend.

As a part of his contract duties, the consulting engineer should supply copies to the owner of his preliminary study and design report, the construction plans and specifications, and a complete detailed operations manual for the plant. The engineer should require the construction contractor to supply complete shop drawings, catalogs, and installation, operation, and maintenance instructions for all equipment.

PLANT OPERATIONS MANUAL

The plant operations manual to be prepared by the design engineer is a most vital and important element in the training program. It must be very detailed and complete. The operations manual should contain descriptions and drawings of normal flow patterns through the plant. It should also include similar information for all alternate modes of operation, and explicit instructions for action to be taken under various emergency conditions, such as electrical power outages of different durations, machinery failures, pipeline stoppages, and the like. The purpose of each unit process should be explained, together with discussions of the relationship of all equipment functions to the process. Some equipment is ordinarily in continuous service and its operation becomes a simple routine. On those rare occasions when it must be shut down for maintenance or repair such operations are less familiar and are often more difficult or complex. It is good to have a written step-by-step procedure to follow for both shutdown and start-up of all equipment and controls of this type. The automatic control system and electrical circuits must be completely diagramed not only for the benefit of the plant operator, but also for the reference and use of outside electricians, control experts, and manufacturers' representatives who may be called in to render assistance under special circumstances. Again, since the manual mode for alternate operation of fully automated devices is seldom used, it too should be recorded in the operations manual in detailed step-by-step fashion.

The essential methods, observations, and tests for control of plant operations should be spelled out, and the key controls must be clearly differentiated from those tests and procedures which are performed merely for purposes of general information or record keeping. There are always situations in plant operation when there is only enough time to do the bare essentials required to keep the plant on the line, and the operators must know which tasks are vital and those which are nonessential and which can be deferred for the moment. Sampling and laboratory testing should be described in the briefest, clearest manner possible. This will require condensation of the rather lengthy and detailed procedures contained in *Standard Methods*. Some of the necessarily technical language in *Standard Methods* can be translated into the simple, understandable bare essentials for the operator.

In starting a new plant, or in adding new processes to an existing plant, the engineer who designed the plant and is familiar with how it is supposed to operate is probably the only person at this stage in development of plant operations who is in a position to prepare forms for reporting plant flow and operating data and laboratory results, and he should do this job. It is also helpful for the engineer to prepare forms for summarizing all plant data, including operating costs, on a weekly, monthly, or annual basis, or all three.

To illustrate the form which a typical operations manual for an advanced wastewater treatment plant might take, an outline is presented below. This is taken from the Tahoe Project as prepared by Cornell, Howland, Hayes & Merryfield and Clair A. Hill & Associates.

Outline of Typical Operations Manual

TABLE OF CONTENTS

- Secondary Treatment
- Sludge Recirculation Pump Station No. 1
- Recirculation Pump Station No. 2
- Sludge Flow Division Box
- Return Sludge, Control Center
- Raw Sludge and Waste Activated Sludge Pumps
- Waste Activated Sludge Flotation Thickener
- Digesters
- Secondary and Final Effiluent Pump Stations
- Ballast Ponds
- Separation Beds
- Backwash Water Decanting Tank
- Carbon Columns
- Ammonia Stripping
- Recarbonation
- Chemical Application
  - Chlorine
  - Lime (Splitter Box)
  - Polyelectrolyte
    - Separation Beds
    - Other Points
    - Centrifuges
  - Alum
  - Carbon Dioxide (Flue Gas)
- Activated Carbon Treatment
  - Purpose
  - Carbon Transfer
  - Carbon Regeneration
    - Control of Carbon Regeneration
    - Apparent Density Test for Carbon
    - Operation of Carbon Regeneration Furnace
      - Startup
      - Shutdown
      - Malfunctions
      - Carbon De-fining
- Solids Handling
  - Raw and Biological Sludge
    - Chemical Sludge
      - Alum
      - Lime
- Emergency Operation
  - Electric Power Outage
  - Natural Gas Off

PROCESS

- General
- Prechlorination
- Screening and Sewage Grinding
- Chemical Feed
- Primary Sedimentation
- Sludge Withdrawal
- Secondary Treatment
  - General
  - Aeration Basins
  - Diffused Air Aeration
  - Mechanical Aeration
  - Return Activated Sludge
  - Waste Activated Sludge

Chemical Treatment and Phosphate Removal
- General
- Chemicals
- Rapid Mix
- Flocculation
- Chemical Clarifier
- Recarbonation (Alternate)

Nitrogen Removal
- Ammonia Stripping
- Primary Recarbonation
- Reaction Basin
- Secondary Recarbonation

Filtration
- Chemicals
- Separation Beds
- Backwashing
- Decanting Tank

Carbon Adsorption

Disinfection

Granular Activated Carbon Regeneration

Sludge Incineration
- Waste Activated Sludge Thickening
- Primary and Waste Activated Sludge Dewatering
- Burning
- Ash Disposal

Lime Recalcining
- Lime Mud Thickening
- Dewatering Lime Sludge
- Recalcining
- Lime Reuse
- Makeup Lime Use

CONTROL OF PLANT OPERATION

Primary Treatment
- Lime Application and *p*H
- Sludge Withdrawal

Secondary Treatment
- Mixed Liquor Solids
- Load Factor
- Sludge Volume Index (SVI)
- Dissolved Oxygen (DO)
- Return Sludge

Chemical Treatment
- Flocculation Basin, *p*H
- Recarbonation (Alternate)

Ammonia Stripping
- Influent *p*H
- Air to Water Ratio
- Recarbonation (Two-Stage)

Filtration

Carbon Adsorption

Carbon Regeneration

Sludge Incineration
- Solids Content of Thickened Sludge
- Solids Content of Dewatered Sludge

Lime Recalcining
- Solids Content of Dewatered Sludge
- CaO Content of Recalcined Lime

LIST OF FIGURES

Secondary and Chemical Clarifiers—Sludge Withdrawal Pipes
Sludge Recirculation Pump Station No. 1—Normal Operation
Sludge Recirculation Pump Station No. 1—Secondary Sludge Pump No. 1. Out of Service
Sludge Recirculation Pump Station No. 1—Secondary Sludge Pump No. 2 Out of Service
Sludge Recirculation Pump Station No. 1—Secondary Sludge Pump No. 3 Out of Service
Sludge Recirculation Pump Station No. 1—Backflush Sludge Lines from Sec. Clarifier No. 2 and Chemical Clarifier
Sludge Recirculation Pump Station No. 1—Pumping Down Sec. Clarifier No. 2
Sludge Recirculation Pump Station No. 1—Pump Down Chemical Clarifier
Sludge Recirculation Pump Station No. 1—Pumping Down Flash Mix and Floc Basins
Pumping Recirculation Pump Station No. 2—Normal Operation
Sludge Recirculation Pump Station No. 2—Sludge Thickener Out of Service
Sludge Flow Division Box—Normal Operation
Sludge Flow Division Box—Primary Clarifier No. 1 Out of Service
Sludge Flow Division Box—Primary Clarifier No. 2 Out of Service
Sludge Flow Division Box—Flotation Thickener Out of Service
Return Sludge, Control Center
Primary Sludge & Waste Activated Sludge Pumps—Normal Operation
Primary Sludge and Waste Activated Sludge Pumps—Centrifuges Out of Service
Primary Sludge and Waste Activated Sludge Pumps—Sludge Transfer Pump No. 1 Out of Service
Primary Sludge and Waste Activated Sludge Pumps—Sludge Transfer Pump No. 2 Out of Service
Primary Sludge and Waste Activated Sludge Pumps—Sludge Transfer Pump No. 3 Out of Service
Primary Sludge and Waste Activated Sludge Pumps—Sludge Transfer Pump No. 4 Out of Service
Primary Sludge and Waste Activated Sludge Pumps—Use of Sludge Transfer Pump No. 2 to Clean Sludge Line from Primary Clarifier No. 2
Waste Activated Sludge Flotation Thickener Secondary and Final Effluent Pump Stations—Normal Operation
Secondary and Final Effluent Pump Stations—No. 3 Water Supply Line Out of Service
Secondary and Final Effluent Pump Stations—Final Effluent Pumps Out of Service
Secondary and Final Effluent Pump Station Quality of Final Effluent Not Good Enough for Export
Secondary and Final Effluent Pump Station—Pumping from 60 mg Emergency Storage Pond to Luther Pass
Ballast Ponds—Normal Operation
Ballast Ponds—Pond No. 1 Out of Service
Ballast Ponds—Pond No. 2 Out of Service
Ballast Ponds—Tertiary Plant Out of Service
Effluent and Pond No. 1 Used for Final Effluent Return from Emergency Pond for Tertiary Treatment
Separation Beds—Filter Cycle
Separation Beds—Backwash Cycle
Separation Beds—Filter to Waste Cycle
Strainer in Connecting Pipeline Between Filters
Backwash Water Decanting Tank—Normal Operation with Decant Cycle Timer in Use
Decanted Water to Ballast Ponds; Lime Sludge to Chemical Rapid Mix Basin
Backwash Water Decanting Tank—Tank Used for Flow Equalization Only—All Backwash Water Flows Through Chemical Clarifier; Decant Cycle Timer Not Used
Backwash Water Decanting Tank—Decant Cycle Timer in Use; Alum Sludge to Primary Tank; Decanted Water to Ballast Ponds

Backwash Water Decanting Tank—Tank Used for Flow Equalization Only; All Backwash Water Flows Through New Primary Tank; Decant Cycle Timer not Used
Carbon Columns—Normal Upflow Operation
Carbon Columns—Upflow Through Carbon Column to Waste
Carbon Columns—Reverse Flow
Carbon Columns—Bypassing Carbon Columns
Ammonia Stripping Tower and Recarbonation Basins—Normal Operation
Ammonia Stripping Tower and Recarbonation Basins—Stripping Tower Only Bypassed; Gravity Flow Through Recarbonation Basins
Ammonia Stripping Tower and Recarbonation Basins—Tower and Recarbonation Basins Both Bypassed
Points of Chlorine Application
Lime Application
Polyelectrolyte (Secondary Flocculant) Feed to Separation Beds
Alum Application
Recarbonation Schematic
Section Through Carbon Column
Control Schematic for Carbon Slurry Pumps
Carbon De-fining Tank
Cargon Regeneration System
Apparatus for Determining the Apparent Density of Granular Activated Carbon
Solids Handling—Primary and Secondary Sludge
Solids Handling—Lime Sludge
Primary Treatment Barminutor
Lime Feed Piping and Mud Thickener Overflow Piping to Primary Clarifiers Nos. 1 and 2
Raw Sludge Pumping
Sludge Wasting Control Systems
Polyelectrolyte Feed Pump Settings
Process Control Diagram
Percent Ammonia Available for Removal by Stripping at 77°F as a Function of *p*H
Phosphate Concentration and *p*H Versus Lime
Dosage
Iodine Correction Curve
Specific Conductance Correction Curve

DIFFERENCES IN OPERATING ADVANCED VERSUS CONVENTIONAL PLANTS

One important difference between the operation of advanced wastewater treatment plants and that experienced historically with conventional secondary plants must be stressed in operator training. That is, the plant shall be operated so as produce the required high-quality reclaimed water at all times and without exception. This kind of operation is accomplished routinely and universally in water purification plants and in electric generation stations. It has also been demonstrated recently that this kind of operation is feasible and practical in advanced wastewater treatment plants. The Water Reclamation Plant of the South Tahoe Public Utility District has operated for two years and all of the water produced has met the high discharge requirements continuously and without exception. Unfortunately, this has seldom been done in secondary treatment plants. Momentary treatment upsets, or even prolonged ones, are the rule rather than the exception in the operation of primary and secondary sewage treatment plants. Many of these upsets are not even known to the plant operator, and are often not detected by the regulatory authorities. In small and medium-sized plants, operator attendance may be regularly scheduled only on an 8-hr day, 5-day week basis. If the plant goes off the line at night, it probably will not be remedied until the following day. If major trouble occurs on Friday afternoon, the plant, or parts of it, may be bypassed until repairs are made on the following Monday. Activated sludge treatment and anaerobic digestion of sludge are two biological processes which are widely used in secondary treatment plants and which are difficult processes to control. It is sad but true that the operation of the great majority of existing secondary wastewater treatment plants is quite erratic and often badly neglected. Once built, many plants are then ignored or their existence almost forgotten. In many, many plants virtually the entire capital investment in sewage treatment facilities is lost for all practical purposes, because the plants were never really operated, and the potential benefits of pollution abatement were never fully realized.

The public is receiving much less value for its pollution control expenditures than it realizes. Certainly, the municipalities and engineers involved must now face this fact and take steps to insure that operator training and compensation are raised to adequate standards to insure the proper and continuous performance of wastewater treatment facilities.

In situations where there is a need to construct an advanced wastewater treatment plant or a water reclamation plant, part-time attendance and poor operation cannot be tolerated, or the whole purpose of the plant and the investment it entails may be lost or defeated. Good operation is essential to successfully accomplish the purpose of advanced treatment. This is not a difficult task, but one which requires some interest, attention, and effort, and it cannot be done without paying sufficient wages to recruit and retain people who can acquire the necessary knowledge and develop the essential skills to get the job done.

The cost of clean water is quite reasonable. Sewer service charges are the lowest of all utility bills. The public is demanding that water pollution cease. The cost of providing good wastewater treatment plant operation is only slightly greater than that for providing poor or inadequate operation, and plant operation is a very small part of the total cost of providing sewer service. Good operation can be provided quite simply by devoting the necessary attention to financing plant operations and properly training plant operators.

The benefits resulting from improved water quality will much more than offset the cost difference between providing mediocre and good plant operation. It is not possible to overstress the importance of good plant operation and the part it plays in the overall success of water pollution control and water reclamation projects.

References

1. "Elementary Mathematics and Basic Calculations," *Water and Sewage Works.*
2. *Manual for Sewage Plant Operators,* Texas Water & Sewage Works Assoc., Austin, Texas.
3. *Operation of Wastewater Treatment Plants,* Manual of Practice No. 11, WPCF, Washington, D.C.
4. "Operator Short Course," *Water & Wastes Engineering.*
5. *Standard Methods for the Analysis of Water, Sewage, and Industrial Wastes,* American Public Health Assoc., New York.
6. "The Operation of Wastewater Treatment Plants," *Public Works,* 1970.
7. *Wastewater Treatment Plant Operator Training Course Two,* Publication No. 14, WPCF, Washington, D.C.

Selecting and Combining Unit Processes to Obtain the Desired Water Quality

AVAILABLE UNIT PROCESSES

In preceding chapters the presently available unit processes for advanced wastewater treatment have been described. These unit processes were placed in two general categories: those which have been successfully demonstrated on a full plant scale, and those which show promise for future development. Such advanced wastewater treatment processes are listed in the respective two categories in Table 13–1, along with certain conventional treatment processes with which they may be used in combination.

Table 13-1 Available Unit Treatment Processes.

Principal Purposes of Unit Processes	*Unit Processes Demonstrated at Plant Scale*	*Potential Unit Processes*
Grit removal	Grit chambers	
Removal or grinding of coarse solids	Bar screens Comminutors	
Odor control	Prechlorination Ozonation	

Table 13-1 (Continued)

Principal Purposes of Unit Processes	*Unit Processes Demonstrated at Plant Scale*	*Potential Unit Processes*
Gross solids-liquid separation; BOD reduction	Plain primary settling	Flotation
Gross removal of soluble BOD and COD from raw wastewater	Biological treatment	Carbon adsorption
Removal of oxidized particulates and biological solids	Plain secondary settling	
Decomposition or stabilization of organic solids; conditioning of sludge for dewatering	Anaerobic sludge digestion	
Improve sludge dewatering characteristics	Anaerobic digestion Thickening: Gravity Flotation Elutriation Heat treatment Ash conditioning Chemical conditioning: Chlorine Alum Lime Polymers Iron salts	Freezing
Preparing organic or chemical sludge for disposal or further treatment	Dewatering organic or chemical sludge: Air drying Centrifuging Vacuum filtration Coil septum Fabric septum Filter press	
Ultimate sludge disposal	Incineration Multiple-hearth Fluidized bed Land disposal Injection Recovery and reuse of chemical sludges	Incorporation into useful products (building materials, fertilizers, etc.) Land reclamation

Table 13-1 (Continued)

Principal Purposes of Unit Processes	*Unit Processes Demonstrated at Plant Scale*	*Potential Unit Processes*
Removal of colloidal solids and turbidity from wastewater	Chemical treatment, sedimentation, and mixed-media filtration: Alum Lime Polymers Iron salts	
Phosphorus removal	Chemical coagulation, flocculation, and settling: Lime Alum Iron salts	Combined biological-chemical treatment; Biological uptake; Selective ion exchange
Nitrogen removal	Ammonia stripping	Breakpoint chlorination for ammonia removal; Selective ion exchange; Microbial denitrification
Recarbonation *p*H adjustment, control of corrosion and scaling; calcium recovery		Sulfuric acid treatment
Removal of suspended and colloidal materials; protection of granular carbon beds or ion exchange beds from fouling or plugging	Mixed media filtration Dual media filtration	Moving bed filter Microscreening (non-chemical solids)
Removal of dissolved trace refractory organics—MBAS, COD, BOD, color, odor, etc.	Granular activated carbon adsorption: Upflow packed Upflow expanded Downflow series beds	Powdered carbon adsorption: Chlorination Ozonation Electrolytic treatment
Disinfection; bacteria and virus inactivation	Chlorination	Electrolytic treatment Ozonation
Removal of dissolved salts; reduction of total dissolved solids content		Demineralization Ion exchange Distillation Reverse osmosis Electrodialysis

FACTORS AFFECTING SELECTION OF UNIT PROCESSES

Effluent Quality Required

The first consideration in the selection of unit processes is the degree of treatment required at present and that which may be required in the future. General water quality criteria for wastewater discharge or for various municipal, industrial, recreational, and agricultural uses of water are discussed in the first chapter of this book. Once the required water quality is established, one must then select unit processes to accomplish the desired result. Obviously, for almost every situation, there are a number of combinations of unit processes which will satisfy conditions. Known and possible future water quality requirements must be kept in mind so that unit processes selected for the original plant can be expanded or supplemented readily to meet future needs.

Reliability

The simplicity and reliability of unit processes and combinations thereof are top-priority factors. While simplicity and reliability are not synonymous, they are very closely related. The simpler a process is to operate and control, the more likely it is to operate successfully and continuously. Regardless of how highly efficient a process may be when operating at its very best under rigidly controlled conditions, if it is sensitive to minor changes in flow, temperature, or applied water quality, or if it requires constant expert control and supervision, then it is not likely to operate as intended all of the time in practical plant application. Processes of this type may make excellent research or laboratory exercises and may be instrumental in developing useful new treatment concepts, but they may be ill-suited to direct adaption to plant-scale use. Simplicity also adds to reliability by making it easier to design and engineer the structures and equipment needed to carry out the process.

The state of development of a process, particularly the number of full-scale plants from which engineering design and equipment information, actual operating results, and cost data can be obtained, has a great influence on the degree of reliability which may be expected of it. Obviously, if new approaches are not tried on a plant scale, little progress will be made. However, the engineer and client involved should enter such a project aware of the degree of uncertainty involved. The acceptable risk level must be determined individually for each project.

The operation of conventional secondary treatment plants is often notably erratic. Advanced wastewater treatment processes which follow secondary ones must take this fact into account. They must be able to handle these variations in performance and still produce an effluent of uniform quality.

The first tests of any new treatment method are usually made in the laboratory. The next step is pilot plant operation on a continuous flow basis. Pilot plant work gives a good insight into the results to be expected from liquid processing. Unfortunately, it does not always adequately evaluate sludge problems or problems associated with recycled process streams or return solids. Only a few advanced wastewater treatment processes have been tested on a plant scale. It is only at plant scale that the real enginering design problems and the practical operation problems are faced for the first time. The transformation of pilot plant data gathered by biologists, engineers, chemists, and laboratory technicians into workable engineering designs is a difficult and demanding task, and one not to be taken lightly or for granted. It is here, in the hands of the sanitary engineering designer, that the project either fails or succeeds. The success depends not only on how well he selects and combines the unit processes, but more importantly how well he selects and specifies the performance of equipment, and how well the excruciating details of design are executed. Actually, the development of this engineering design know-how forms the great gap in the development and application of new and better methods for wastewater treatment. There is a great backlog and wealth of information and scientific research which cannot be used until it is applied to actual working plants. This transformation requires the design of new structures and equipment, and the provision of sufficient safety and flexibility in the design to allow for any contingencies which may not be apparent or solvable at pilot scale.

As already pointed out, the problems involved in going from pilot to full scale are difficult, so that it is sensible to take full advantage of all prior knowledge. There are many unknowns still remaining upon completion of pilot-scale experiments which become readily discernible upon completion and operation of the full-scale plant. For maximum reliability, the designers of future plants must take advantage of the prior art, modifying proven designs as necessary to meet different conditions as to size and other common design variables.

Sludge Handling

Of major importance in process selection are the circumstances related to the handling and ultimate disposal of the sludge produced. In locations where there are available large, remote areas of land, almost any kind of sludge, wet or dry, stable or decomposing, can be, and is, disposed of by hauling or pumping to these land disposal sites. In many places this method for sludge disposal probably will not be tolerated indefinitely, but might suffice for the time being. Ocean dumping has long been an easy way to evade the knotty problems involved in proper sludge disposal. However, the nuisances which have been created and the damages wrought to beaches and coastal waters have aroused the public to the point where this method is now in almost universal public disfavor.

All of the good methods for sludge disposal involve dewatering of the sludge. As pointed out previously, the ease with which sludge may be dewatered is a prime factor in unit process selection. There are many alternate ways to process the liquid component of wastewater to secure the desired result at about equal costs, but there are very few ways to satisfactorily and economically dewater sludge. In certain wastewaters, dewatering of mixtures of organic-chemical sludges may be satisfactory, but care must be taken to check this out before designing a full-scale plant. Favorable pilot plant tests are a prerequisite.

Another approach which has been used successfully is to keep all organic sludges entirely separate from all chemical ones. Then, conventional equipment used for either of these types of sludge can be installed. Pilot tests are still highly desirable, even with this approach.

Heat treatment or anaerobic digestion of organic-chemical sludges may condition these mixtures to make possible use of a wide range of dewatering equipment. It is also possible that chemical conditioners will accomplish this same end, but more research and development work is needed before this can be realized.

Process Compatibility

A fourth consideration in process selection is compatibility with other unit processes used in the overall treatment scheme. The possible effects of waste streams or recycled solids is also very important.

The most favorable point for addition of chemical coagulants and flocculants may be influenced by the form in which phosphorus is present. Raw sewage in primary tanks, for example, usually contains polyphosphates which are later broken down to orthophosphate by biological treatment as discussed earlier. Polyphosphates in relatively high concentrations (more than 1 mg/l of P_2O_5) are capable of interfering with coagulation and sedimentation. Orthophosphate compounds produce no interference with normal coagulation and sedimentation. If the amount of polyphosphate interference in the primary stage is great, then, obviously chemical addition should follow reversion of this material to orthophosphate.

The optimum *p*H for various unit processes is a factor which influences their compatibility, and may affect placement of processes in the plant flow sequence. There are many examples of this point. Various chemical coagulants operate best within certain *p*H ranges. In this case, there may be a choice—either change the *p*H or change the chemical used. Also it is possible to broaden the effective *p*H range of chemicals by use of coagulant aids, such as activated silica or polymers. Biological processes may be adversely affected by *p*H values above 9.3, although it is well to keep in mind that activated sludge cultures have great buffer capacity and it may be possible to allow the *p*H of the influent to activated sludge basins to exceed 9.3 so long as the *p*H in the aeration tank itself does not exceed this value. Phosphorus removal by lime requires raising the *p*H to values of 9–11. Ammonia stripping also is most efficient at *p*H values of 10.5–11. Calcium reaches minimum solubility at a *p*H near 9.3. Carbon adsorption of organics is best at low *p*H values, and is very poor at values above 9.0. Chlorination is more effective at low *p*H values, at low turbidities, and in waters with low chlorine demand. Granular activated carbon adsorption is favored by good clarification of the applied water. Granular activated carbon will adsorb chlorine. With information on process interrelationships and *p*H effects, then, a logical process sequence for maximum reliability and effluent quality would be biological treatment followed by advanced treatment, or the so-called tertiary sequence. Although the costs may exceed those of approaches combining certain steps, there are at least three advantages to this arrangement which must be given due consideration: the *p*H is favorable for biological treatment; the organic sludges are kept separate from the chemical types, which usually minimizes sludge dewatering problems; and the polyphosphates in the raw wastewater are converted to orthophosphate before chemical coagulants are applied, thus avoiding potential interference with chemical coagulation.

In considering compatibility, attention must be given to the point in the overall process at which centrate, filtrate, backwash water, scrubber water, and other plant process water and recycle streams are introduced. It is wise to provide more than one point of return for some of these streams, particularly to take care of those times when certain plant units are temporarily out of service.

Costs

The final factor in process selection is indeed important, and that is capital and operating costs. It is obviously and patently false economy to use a process which will not do the job because it is cheap to install and operate. In those cases where only a low risk level is acceptable, if the designer has a choice between a proven process with a high degree of reliability and a process which may be cheaper, but is of unknown or questionable reliability, then prudence is on the side of the more costly but workable process. One special word of caution is appropriate with regard to the attractive and tempting possibility of combining various processes into a single process or into a single structure. By reducing the total number of structures, there are obvious savings in construction costs. Actually, this initial cost saving is usually the only advantage of combining processes, and there may be many disadvantages. The history of water and wastewater treatment is replete with failures to satisfactorily combine two or more processes. Yet, some of these mistakes are perpetuated in present-day designs, and new unworkable combinations may be devised. This is not to say that some of these efforts may not end in success, but only to inject a word of caution and to point out some of the inherent weaknesses in making one process out of two or more. First of all, there is invariably a loss of control or flexibility in operation by combining two processes. A change which benefits one of the processes may interfere with or reduce the efficiency of the other. Also, it is much more difficult to relate cause and effect, which is necessary to make treatment adjustments, because of the greater complexity introduced by the combination. There are many specific examples which can be presented.

Upflow solids contact basins of the sludge blanket type combine the four processes of rapid mixing, flocculation, settling, and solids recycling in a single unit. They have been applied successfully to treatment of a constant flow of water (principally well waters) which are of constant physical and chemical quality, at a considerable savings over the use of four separate operations. However, these sludge blanket basins are often not stable when operated at varying flow rates, or with waters which have significant variations in chemical composition or physical characteristics, such as temperature or organic content. It is difficult to change the point of chemical addition in this type of basin if it becomes necessary. Also with changes in flow rate, or chemical composition or physical character of the applied water, or the accumulation of organics in the sludge blanket, the chemical dosages required to maintain a sludge blanket of the proper density change so frequently and rapidly as to present an impossible operating condition, and the basin fails to function as a settling device. Wastewaters possess properties which make operation of sludge blanket clarifiers difficult. The experiences cited earlier indicate that this type of equipment can be used satisfactorily for wastewater treatment if the overflow rates are the same as for a conventional horizontal flow clarifier, and if the sludge is removed from the basin as rapidly as it reaches the bottom of the tank. Separate rapid mixing, flocculation, and settling basins with provisions for external sludge recirculation are much more satisfactory for treatment of wastewater. Mutual interference is avoided, and complete flexibility of operation and control is afforded.

The difficulties and high costs which may be involved in dewatering organic-chemical sludge mixtures in some wastewaters were discussed previously. Combining high lime or alum treatment with primary settling with the elimination of a separate chemical clarifier, or even going further and eliminating biological treatment, offers potential savings which are quite substantial. The possibilities certainly warrant investigation in each instance to see if there is a sludge dewatering problem, and, if so, if there is an economical solution to it. This should be done on a pilot plant scale before undertaking the building of a full-scale plant.

Combining chemical treatment with activated sludge treatment for phosphorus removal also presents some practical difficulties at full plant scale which should not be overlooked, even though there may be ways to solve the problems. Addition of lime or alum to aeration tanks secures excellent phosphorus removal. In some locations it is reported that the inert chemical sludge accumulates and selectively displaces the culture of activated sludge organisms over a period of a few weeks, and normal biological treatment ceases. Because of the differences in specific gravity between the chemical and biological sludges it may be possible to separate them and return or waste them as necessary to control the process, but this remains to be demonstrated on a plant scale.

Combination downflow carbon contactor-filters are used successfully in water treatment plant practice where there are few suspended solids in the applied water. However, the carbon contactor-filter is a surface filter, not an in-depth type, and is subject to the same limitations discussed earlier for any other surface filter in filtering wastewater. The use of granular carbon as a contactor-filter has all of the same disadvantages of other surface filters plus the additional one of being a much deeper bed, and therefore more difficult to properly clean by backwashing than a shallower type. Also, the use of carbon as a filter to remove suspended and colloidal matter from the water is likely to blind the carbon pore openings and require removal of the carbon from the bed before its full adsorptive capacity is exhausted.

Flocculation and settling of organics and lime sludge in the same basin poses another problem if the lime is to be recalcined and reused. Calcium carbonate has been recalcined and organics incinerated successfully in a single furnace. However, this means that the recalcined lime contains considerable inert ash. There are two choices, the ash can be recycled with the recalcined lime, or some means can be devised to separate the lime from the ash. If the ash is recycled with the lime it becomes wet again and there are extra costs for fuel for repeated drying of the ash on subsequent cycles. It is necessary to waste about 25–35 percent of the recalcined lime on each cycle of use to rid the system of phosphates, and a similar portion of the recycled ash would automatically be wasted in the same operation. This would be an acceptable way to operate, and the decision as to whether to settle and incinerate organics and chemical sludge together or separately can be resolved by a rigid cost analysis including, of course, all capital and operating costs. Whether to attempt to separate the ash from the reclaimed lime as opposed to recycling the ash with the lime also is strictly a matter of comparative costs, assuming that the separation can be made successfully.

These examples should serve to illustrate some possible problems which may be encountered when attempting to combine two or more unit processes into a single unit. Combining processes usually complicates operation and control and decreases reliability, while the designer should be seeking to simplify operation and control, provide maximum flexibility of operation, and insure greater reliability. There is no point in combining two perfectly good unit processes into one poor one. Because of the potential economic benefits, efforts to determine workable combinations should certainly continue.

One way to reduce capital costs for construction of advanced treatment plants is to introduce storage of wastewater at some point in the system. This should be done after the water has received sufficient treatment so as not to create a nuisance. Sufficient storage should be provided to allow the sections of the plant which follow the storage to operate at the average rate for the design maximum day rather than at the peak hourly rate for the maximum day. A rough rule of thumb for separate sanitary sewer systems is to build at least 1.5–2.0 mg of storage for each 10 mgd of plant capacity. The exact requirements vary for each individual collection system, of course, and the rule of thumb does not apply to systems receiving storm water as well as sanitary sewage.

Special Factors Associated with Future Water Reuse for Potable Supply

In the future, when situations develop where the only possible solution to especially severe public water supply shortages is to recycle domestic wastewater, there will be special consideration of reclaimed water quality related to public health which will affect unit process selection, as well as the combination of these processes, which make up the total water reclamation plant.

Present concern is not so much as to whether reclaimed wastewater *can* be used as a source of drinking water supply, but rather whether it *should* be so used. Indications are that water quality suitable for direct potable use can be produced from wastewater under certain favorable conditions when there is sufficient justification, but as discussed in Chapter 1, this justification does not now exist in the United States.

At Windhoek, South Africa, which is located in an extremely water-short area, about one third of the municipal water supply is now obtained from reclaimed wastewater.

If reclaimed water had to be used for potable purposes today, several additions and modifications could be made to increase the safety of the water from a public health standpoint. For example, at Windhoek, maturation, or holding ponds, are a part of the process. Holding water for a period of several days allows the natural die-away of bacteria, virus, and other disease-producing organisms to occur. Perhaps the most effective barrier to possible disease transmission from water reuse would be double chlorination with time for intermediate contact with the reclaimed water. It is only reasonable to fully utilize this tool in processing wastewater, because, in the final analysis, all public water supplies now obtained from surface sources are dependent upon adequate chlorination as the principal means of protecting the public health.

The above discussion is based on the premise that only about one third of the total supply would be obtained from this source. With a completely closed loop and with almost 100 percent of the supply coming from recycled water, there is a problem with a buildup of dissolved salts which can only be removed by processes capable of complete demineralization of the water. This is a whole new area of water treatment and will not be discussed further in this book.

THE TAHOE DEMONSTRATION PROJECT

Various parts and aspects of the Tahoe Process for water reclamation have been discussed throughout the text. The purpose of further discussion here is to present summaries of (1) the complete process, (2) design data, (3) performance records, (4) operating experience, (5) complete breakdown of actual construction costs, and (6) detailed actual costs for operation and maintenance of the plant.

The Process

A flow diagram for the complete process is shown in Figure 13–1. The plant will process 7½ mg/day of wastewater. Basically, the treatment is divided into two parts—liquid processing and sludge handling. The first two steps in liquid processing are conventional treatment, including primary treatment, or solids separation; and secondary treatment by biological oxidation. There are five additional steps, all in the nature of advanced waste treatment—chemical treatment and phosphate removal, nitrogen removal, mixed-media filtration, activated carbon adsorption, and disinfection. The nitrogen removal facilities are the only part of the plant which is still in the experimental stage. The ammonia stripping tower has a capacity of only 3.75 mgd.

There are three solid materials to be handled: (1) primary and secondary sludge, (2) spent granular carbon, and (3) lime sludge. The sewage sludges are incinerated to insoluble, sterile ash. The spent granular carbon is thermally regenerated and reused, and the lime sludge is recalcined and reused. Figure 13–2 is an exterior view of the incinerator building.

Primary treatment takes place in two settling tanks of conventional design. Biological treatment is accomplished in aeration tanks of two different designs. One has plug-flow diffused air, and the other, mechanical aeration plus diffused air. The first step in the advanced

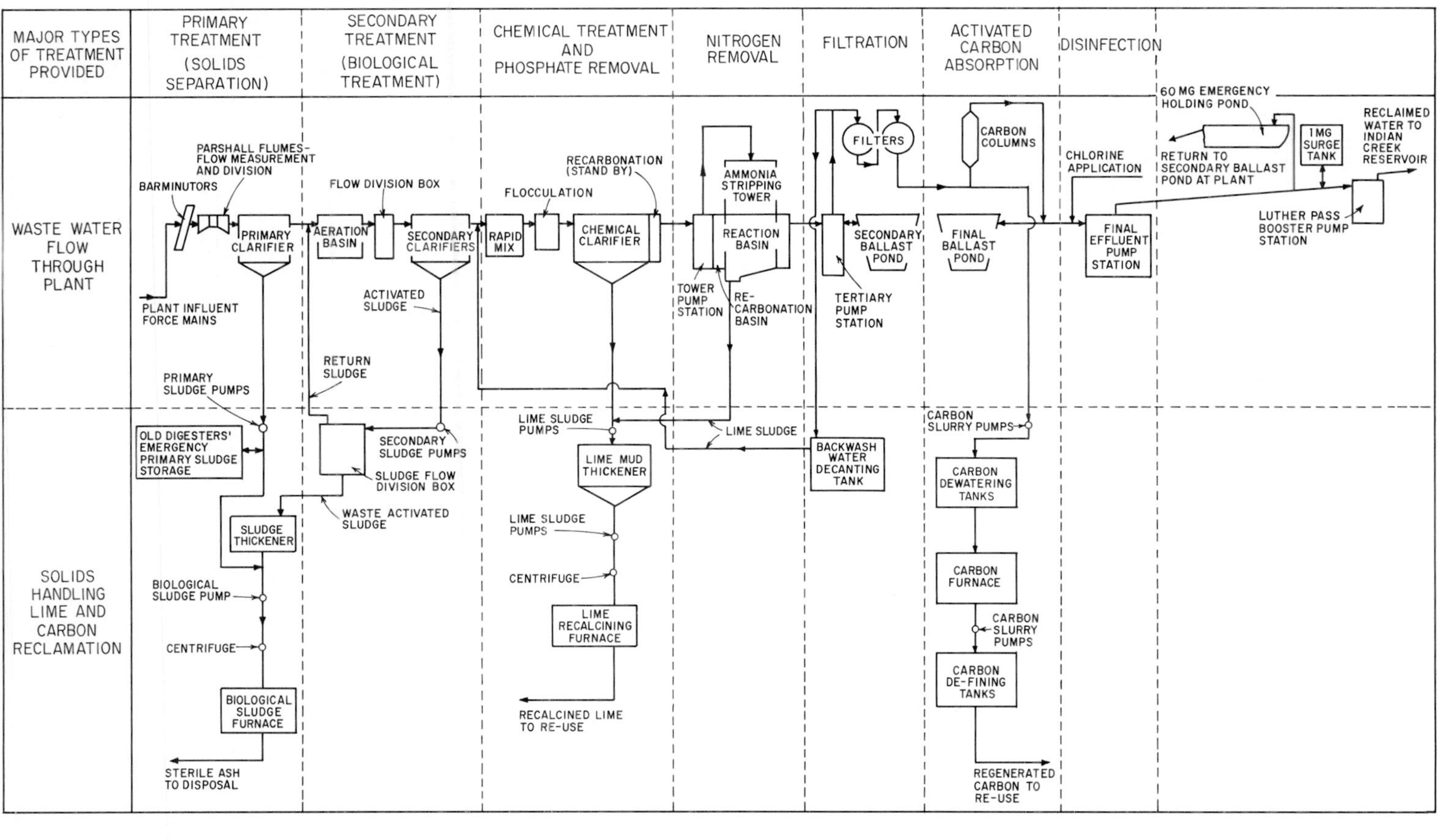

Figure 13–1 Schematic flow and process diagram—South Tahoe Public Utility District Water Reclamation Plant. *(Reprinted from Water & Wastes Engineering, 6:4 (Apr., 1969),* p. 36. *R. H. Donnelley Corp.)*

Figure 13–2 Incinerator building. *(Courtesy Cornell, Howland, Hayes & Merryfield)*

waste treatment consists of the addition of lime, followed by flocculation and clarification of the water by settling. This treatment removes most of the phosphates present and raises the *p*H to a level high enough to convert ammonium ion to ammonia gas. The lime-treated water is then pumped to the nitrogen removal tower where ammonia is stripped from the water by circulation of large quantities of air and droplet formation in the tower. The stripping tower effluent is collected in a catch basin beneath where the water is stabilized by means of adding carbon dioxide gas in two stages of treatment. The first-stage recarbonation drops the *p*H to 9.3 for maximum calcium recovery. The second stage of recarbonation drops the *p*H to 7.0–8.5 as desired. The stabilized water then flows to two ballast ponds in series. The ponds provide storage for filter backwash water and equalize flows to the remainder of the plant. Water is pumped from the pond to the tertiary building where it passes first through mixed-media filters and then the carbon columns. When the filters become plugged with material which has been removed from the wastewater, they are automatically taken off the line, backwashed, and restored to service. The waste backwash water is slowly returned for reprocessing in the plant. The filtered water flows under pressure to the carbon columns. There are 8 carbon columns each containing about 22 tons of granular activated carbon. The carbon column effluent is colorless, odorless, sparkling clear, and low in dissolved organics. The chlorine demand of the treated water is low, which permits very effective use of chlorine for disinfection.

Design Data

The principal design criteria are presented in Table 13-2.

Table 13-2 Summary of Design Data for the South Tahoe Water Reclamation Plant.

Item	*Amount*
Plant design average flow	7.5 mgd
Peak flow rate (except as noted below)	15.0 mgd
Peak flow rate (filters and carbon columns)	8.2 mgd
Maximum hydraulic rate	20.0 mgd
Plant design BOD (summer)	325 mg/l
Plant design BOD (winter)	250 mg/l
Plant suspended solids (summer)	200 mg/l
Plant suspended solids (winter)	150 mg/l
Water temperature (summer)	17°C
Atmospheric pressure (elevation 6,300 ft)	11.6 psi
Primary clarifier No. 1	
Surface area	2,350 sq ft
Flow	2.7 mgd
Overflow rate	1,150 gpd/ft^2
Primary clarifier No. 2	
Surface area	7,850 sq ft
Flow	4.8 mgd
Overflow rate	610 gpd/ft^2
Aeration basins 1, 2, and 3, plug flow	
Flow	2.7 mgd
Volume	115,000 cu ft
Detention	115,000 cu ft
BOD loading	50 lb/1,000 cu ft
Aeration basins 4 and 5, complete mix	
Flow	4.8 mgd
Volume	137,000 sq ft
Detention	5 hr
BOD loading	70 lb/1,000 cu ft

Table 13-2 (Continued)

Item	*Amount*
Secondary clarifier No. 1	
Surface area	2,830 sq ft
Overflow rate	700 gpd/ft^2
Secondary clarifier No. 2	
Surface area	7,850 sq ft
Overflow rate	700 gpd/ft^2
Sludge recirculation	
Secondary clarifier No. 1	1,380 gpm
Secondary clarifier No. 2	3,840 gpm
Sludge	
Primary, dry solids	8,130 lb/day
Waste activated, dry solids	9,760 lb/day
Rapid mixer, mechanical	30 sec
Chemical feed equipment	
Gravimetric lime feeders and slakers	
Recalcined lime	1,500 lb/hr
Makeup lime	1,500 lb/hr
Liquid alum feeders	
Two, each	50 gph
Polymer solution feed pumps	
Four, each	50 gph
Chemical storage	
Recalcined lime	35 tons
Makeup lime	35 tons
Liquid alum	10,000 gal
Flocculation chamber	
Air mix	4.5 min
Chemical clarifier	
Flow	7.5 mgd
Surface area	7,850 sq ft
Overflow rate	950 gpd/ft^2
Lime sludge pumps	
Centrifugal	450 gpm
Progressive cavity displacement	100 gpm
Lime mud thickener	
Flow	450 gpm
Surface overflow rate	1,000 gpd/ft^2
Surface dry solids loading	200 lb/day/ft^2
Thickened sludge solids	8–20%
Ammonia stripping tower, cross-flow	
Wastewater flow	3.75 mgd

Fill (packing):	
Area	900 sq ft
Height	24 ft
Splash bars	3/8 x 1-1/2 in.
Spacing	
Vertical	1.33 in.
Horizontal	2 in.
Recarbonation, two-stage, with intermediate settling	
Flow	7.5 mgd
Carbon dioxide compressors	
No. 1	500 cfm
No. 2	1,000 cfm
No. 3	950 cfm
Reaction basin	
Detention	30 min
Surface overflow rate	2,400 gpd/ft^2
Ballast pond capacity	
No. 1	1.0 mg
No. 2	1.5 mg
Pumps to tertiary plant	
No. 1	1,900 gpm
No. 2	3,800 gpm
No. 3	4,200 gpm
Surface wash booster	500 gpm
Mixed media filters	
Flow	8.2 mgd
Units, 3 sets of 2 series beds	
Hydraulic loading	5 gpm/ft^2
Backwash rate	15 gpm/ft^2
Area each bed	380 sq ft
Surface wash flow	0.6 gpm/ft^2
Waste backwash water receiving tank	
Capacity	80,000 gal
Carbon columns (8), upflow countercurrent	
Flow	8.2 mgd
Carbon volume, each column	1,810 cu ft
Carbon depth, effective	14 ft
Contact time	17 min
Hydraulic loading	6.5 gpm/ft^2
Chlorination equipment	
Three feeders, each	2,000 lb/day
Carbon regeneration furnace, 6-hearth, 54-in. diameter, gas-fired	
Capacity, dry carbon	6,000 lb/day

Table 13-2 (Continued)

Item	*Amount*
Sludge dewatering equipment, concurrent flow centrifuges, 24 x 60 in.	
Organic sludge	
Number	2
Capacity, each, dry solids	450 lb/hr
Lime sludge	
Number	1
Capacity	1,650 lb/hr
Sludge incineration furnace, 6-hearth, 14 ft-3 in. diameter, gas-fired	
Capacity, dry solids	900 lb/hr
Lime recalcining furnace, 6-hearth, 14 ft-3 in. diameter, gas-fired	
Capacity, dry CaO	10 tons/day

Plant Performance

The quality of the reclaimed water is very high. Table 13-3 lists typical values for key quality parameters.

Table 13-3 Typical Quality of Reclaimed Water.

Description	*Average Value*
BOD	<1.0
COD	10
Suspended solids mg/l	0
Turbidity (JU)	0.3
MBAS (mg/l)	0.10
Phosphorus (mg/l as P)	0.06
Color (units)	<2
Odor	none
Coliform bacteria (MPN/100 ml)	<2.0

Table 13-4 affords a comparison of performance of the secondary treatment only, and the advanced wastewater plant process at Tahoe.

Details of water quality at various stages of treatment are given in Table 13-5.

One question frequently raised about the safety of reclaimed water concerns its virus content. Since data available in this area is meager, *any* information is helpful. In two summers of sampling at Tahoe, no virus have been recovered from the chlorinated reclaimed water. The

Table 13-4 Comparison of Removals by Secondary Treatment and Tahoe Process.

Parameter	PERCENT REMOVAL *Secondary Treatment*	*Tahoe Process*
Suspended Solids	89	100
Turbidity	96	99.8
BOD	90	99.8
COD	86	96
MBAS (detergents)	93	98
Coliform bacteria	95	99.99+
Phosphorus	52	99.6
Color	incomplete	100
Odor	incomplete	100

Table 13-5 Water Quality at Various Stages in Treatment.

				EFFLUENT			
Quality Parameter	*Raw Waste-water*	*Primary*	*Secon-dary*	*Chemical clarifier*	*Filter*	*Carbon*	*Chlorinated final*
BOD (mg/l)	300	100	30		3	1	0.7
COD (mg/l)	280	220	40		25	10	10
SS (mg/l)	230	100	26	10	0	0	0
Turbidity (JU)	250	150	50	10	0.3	0.3	0.3
MBAS (mg/l)	7	6	2.0		0.5	0.10	0.10
Phosphorus (mg/l)	12	9	6	2	0.10	0.10	0.10
Coliform (MPN/100 ml)	50 million	15 million	2.5 million		50	50	< 2.0

Table 13-6 Virus Sampling (1969), South Tahoe Public Utility District Water Reclamation Plant.
Virus Recovered (PFU)[1]

Date	*Primary effluent*	*Secondary effluent*	*column effluent (U)*	*Final effluent (C)*	*Sample size (l)*[2]
May 29	3	0	1	0	1 *P, S, U, C*
June 5	—	—	0	0	2 *P, S, U, C*
June 12	—	—	0	0	2 *P, S, U, C*
Aug. 20	3	18	NRD[3]	0	4 *P* & *S*, 20 *U* & *C*
Aug. 27	—	—	NRD	0	12 C
Sept. 11	—	—	0	0	27 *U*, 19.7 C
Sept. 18	179	14	9	0	4 *P* & *S*, 29 *U* & *C*
Sept. 25	NRD	430	0	0	4 *S*, 32 *U* & *C*
Oct. 2	207	320	0	0	4 *P* & *S*, 32 *U* & *C*

[1]PFU = Plaque forming units.
[2]*P* = primary; *S* = secondary; *U* = unchlorinated carbon column effluent; *C* = chlorinated carbon column effluent = final effluent.
[3]NRD = no reliable data.
Tests performed by the FWPCA Laboratory in Cincinnati.

results for the summer of 1969 are shown in Table 13-6. Viruses were recovered from primary and secondary effluent and in two of seven samples of carbon column effluent, but none from nine samples of reclaimed water. It certainly is not possible to arrive at any definite far-reaching conclusions from such an extremely limited amount of data, but at least the results are all favorable to date.

Many public water supplies are obtained from highly polluted surface water sources. There have been few, if any, major outbreaks of waterborne viral disease from use of properly treated water from these sources. In the case of wastewater reclamation, the pollution is more direct and the natural die-away time for organisms is shorter, but the degree of treatment provided is greater than that found in most water purification plants. Therefore, the good results of the virus tests of reclaimed water are not surprising. However, much more data is needed on the fate of viruses in water reclamation processes to assure the safety of such water for potable use.

Operating Experiences

The 7½ mg/day water reclamation plant has been in continuous operation since March 31, 1968, with the exception of the nitrogen removal tower, which was completed in November of 1968. The plant has been operating without interruption for the entire period. The quality of water is excellent and has exceeded at all times the high standards set by the regulatory agencies. During the first 2 years of operation, the plant processed 1.6 billion gallons of wastewater from which the following materials were removed:

1. 48 tons of detergent (MBAS)
2. 1,500 tons of suspended solids
3. 100 tons of phosphorus
4. 1,700 tons of oxygen-consuming substances
5. All color, odor, and coliform bacteria

All of the solid materials removed from the wastewater have been incinerated at temperatures of 1200–1600°F to insoluble, sterile ash.

About 1,300 tons of lime mud and 170 tons of spent activated carbon have been reclaimed and reused in the treatment process.

Table 13-7 gives water quality data for the month of November, 1969, as compared to regulatory requirements.

The results for November, 1969, are typical for the entire two years of operation. The reclaimed water quality has without exception exceeded the requirements of the export standards.

Table 13-7 Water Quality Data—November, 1969.

Description	Alpine Co.	REQUIREMENTS Lahontan R.W.Q.C.B.[2] (% of time) 50	80	100	Plant performance (% of time) 50	80	100
MBAS (mg/l), less than	0.5	0.3	0.5	1.0	0.19	0.35	0.35
BOD (mg/l), less than	5	3	5	10	1.0	2.5	3.9
COD (mg/l), less than	30	20	25	50	9	10	22
Susp.S. (mg/l), less than	2	1	2	4	0	0	0
Turbidity (JU), less than	5	3	5	10	0.4	0.5	1.3
Phosphorus, (mg/l), less than	no requirements				0.06	0.12	0.27
*p*H (units)	6.5–8.5	6.5–9.0				6.6–8.7	
Coliform, MPN/100 ml[1]	Adequately disinfected	Median < 2.0 Max. no consecutive sample > 23,2			Median < 2.0 no. of consecutive samples > 23,0, none		

[1]All 30 samples collected during November, 1969, were found to be free of coliform organisms.
[2]Lahontan Regional Water Quality Control Board.

There are no regulatory requirements for removal of phosphorus or nitrogen. However, phosphorus is being removed in order to restrict algal growths in Indian Creek Reservoir, which receives the reclaimed water. For the first few months of plant operation, phosphorus concentrations in the reclaimed water ranged from 0.5–2.0 mg/l. Operations have been steadily improved, and currently the final phosphorus content is in the range of 0.05–0.10 mg/l. The phosphorus level has been low enough at all times to secure the desired control of algal growth in the reservoir.

The changes in plant operation which have been responsible for the improved phosphorus removals are as follows:

1. Changing the point of polymer addition from the flash mix inlet to the chemical clarifier inlet. This resulted in a finer floc in the flocculation basin, providing more opportunities for contact between floc particles and phosphorus, and increasing removals by about 0.5 mg/l of phosphorus.
2. Using two-stage rather than single-stage recarbonation, which also improved removal by about 0.5 mg/l of phosphorus.
3. Recycling spent lime mud from the bottom of the chemical clarifier to the rapid mix basin, which produced a drop of about 0.3 mg/l in the residual phosphorus.
4. Adding 1–2 mg/l of alum to the mixed-media filter influent, which removed another 0.5 mg/l of phosphorus at this point.
5. Presently the point is being changed to which the flue gas scrub-

ber water is recycled back to the process, which is expected to result in a final phosphorus content of less than 0.06 mg/l.

Nitrogen removal is not required at Tahoe. As previously mentioned, there is an experimental ammonia stripping tower which was placed in service in November, 1968, a full 8 months after the rest of the plant was placed into service. The tower has a noted capacity of 3.75 mgd, which is one half the design capacity of the rest of the plant. This means, at times, that only part of the total plant flow can be treated in the tower. When operating properly, the full-scale tower has performed as well as the pilot tower. At 75°F, ammonia removals as high as 95 percent have been obtained. At 39°F, maximum removals of 88 percent have been secured. Because of loss of efficiency and freezing problems, it is not practical to operate the tower at ambient air temperatures below 32°F, so that if nitrogen removal is necessary under cold weather conditions, alternate or supplemental methods must be used.

Problems and Solutions

The principal problem encountered in operation of the plant—and fortunately it is a problem which has been solved—is that of sludge dewatering. With the soft water which is being processed in this plant, mixtures of sewage sludge (that is, primary sludge and/or waste activated sludge) and lime sludge are very difficult to dewater, and require the addition of prohibitively costly amounts of polymer for adequate dewatering. The solution to this problem has been to keep all lime and recycled water containing all fresh lime or lime sludge out of the primary and secondary sections of the plant. In other words, *all* chemical treatment *follows* biological treatment. This separate handling of the sewage sludges and the chemical sludges permits economical dewatering and burning.

A second, less serious, problem has been that of calcium deposits in pipelines carrying lime slurry, high *p*H water, or lime sludge. This was anticipated to some extent by installing glass-lined pipe for all pump suction lines carrying lime sludge, and extending suction and drain lines under concrete to the ground surface outside for easy cleaning. However, it has also been necessary to clean pump discharge lines and gravity flow lines carrying lime sludge or high *p*H water by use of cleaning pigs. Provision should be made in the design of all such lines for easy insertion and removal of such pigs.

Calcium carbonate deposition also occurs on all parts of the ammonia stripping tower. At Tahoe, this is a soft scale which can quite easily and quickly be removed by a stream of water from an ordinary garden hose or by gentle contact with a rod or stick. New towers should be designed for ready cleaning access by either of these means, thus solving the problems of scale removal. Unfortunately, at Tahoe, some parts of the tower are not readily accessible for cleaning because of obstructions formed by structural members, and attempts are being made to use chemical cleaners to remove the scale in inaccessible areas.

In treating some types of wastewater, there may be a tendency for a hard scale to form, and this would be much more difficult to remove.

Costs

If all the benefits of water pollution control are considered and balanced against the cost of even the most advanced treatment, there is little question but that clean water saves money rather than costing it. Unfortunately, the benefits do not often accrue to the same parties who must pay the costs for cleanup.

Compared to the costs for water service, telephone, electricity, and other utilities, the cost for sewer service is very low. The total costs for the five steps of advanced waste treatment are about equal to the costs for conventional secondary treatment. In other words, it costs about twice as much to produce reclaimed water as it does to discharge secondary effluent. This cost is still so small that it constitutes a reasonable charge for a utility service, and one that will not be objected to by the average domestic user.

Table 13-8 presents a detailed tabulation of capital costs for conventional and advanced waste treatment plants with a 7.5 mgd design capacity.

All costs for operation and maintenance of the Tahoe plant have been programmed on a computer. The treatment phases have been divided into primary treatment, secondary treatment, organic sludge handling and dewatering, organic sludge incineration and disposal, disinfection, phosphorus removal, lime recalcining, ammonia stripping, recarbonation, filtration, carbon adsorption, and carbon regeneration. For each treatment phase, operational costs such as electricity, labor, repair materials, natural gas, and chemicals are tabulated.

The unit costs are as follows: operations labor, \$5.90 per hour; maintenance labor, \$5.28 per hour; fuel, \$0.054 per therm; and power, $0.0085 per khw plus $2.00 per kw demand. Lime costs are based on a dose of 400 mg/l and a delivered cost of $26 per ton.

Table 13-8 South Tahoe Public Utility District, California, Capital Costs for Conventional and Advanced Waste Treatment Plant 7.5 mgd Design Capacity.[1]

Treatment Phase	*Actual Contract Total Construction Cost Per Phase*[2]	*Estimated National Average Replacement Construction Cost for 1969*[3]	*Estimated Replacement Costs Per MG for 1969*[4]
Conventional Treatment			
Primary	$ 692,000	$ 753,000	$ 19.50
Activated sludge	1,247,000	1,300,000	33.60
Organic sludge[5]	583,000	545,000	14.10
Chlorination	9,000	11,000	0.30
Total, Conventional Treatment	$2,531,000	$2,609,000	$ 67.50
Advanced Treatment			
Nutrient removal			
Phosphorus removal			
Lime treatment	401,000	378,000	9.70
Lime recalcining[6]	552,000	516,000	13.50
Subtotal, Phosphorus Removal	$ 953,000	$ 894,000	$ 23.20
Nitrogen removal[7]	327,000	310,000	8.00
Recarbonation	162,000	152,000	4.00
Subtotal, Nutrient Removal	$1,442,000	$1,356,000	$ 35.20
Filtration	705,000	687,000	17.80
Carbon treatment			
Carbon adsorption[8]	656,000	632,000	16.30
Carbon regeneration	193,000	199,000	5.20
Subtotal, Carbon Treatment	$ 849,000	$ 831,000	$ 21.50
Total, Advanced Treatment	$2,996,000	$2,874,000	$ 74.50
Total Water Reclamation			
Conventional treatment	2,531,000	2,609,000	67.50
Advanced treatment	2,996,000	2,874,000	74.50
Total, Water Reclamation	$5,527,000	$5,483,000	$142.00

[1]Includes all equipment and construction costs. Does not include design costs.
[2]Construction costs are taken from District records of actual contracts awarded for various phases. Contracts for construction were completed at various periods between 1960 and 1968. These costs have not been adjusted to a common year.
[3]National average replacement costs are the actual contract costs, for each year of construction, adjusted to 1969. Adjustments were made using the FWPCA Sewage Treatment Plant Construction Cost Index, Base Year 1957–59 = 100. Estimated 1969 index values were 136.2 for the San Francisco Region and 127.1 for the Nation. The San Francisco Regional Index was used for the South Tahoe area.
[4]Computed from National average 1969 replacement costs assuming capital amortization of total costs at 5% interest for 25 years.
[5]Includes sludge handling, dewatering, incineration, and ash disposal.
[6]Includes lime mud handling, dewatering, and recalcining.
[7]Ammonia stripping.
[8]Includes initial carbon costs to fill all carbon columns.
[9]Federal grants from the USPHS, FWPCA, and EDA financed approximately 46% of the total construction cost.

Table 13-9 is taken from the computer cost run for the period of February 1969 through January 1970.

Table 13-9 South Tahoe Public Utility District Water Reclamation Plant Operational Cost Analysis for the Year February 1969 through January 1970.

	Units	*Costs at 7.5 mgd Flow*
Operating Cost for		
Conventional Waste Treatment		
Primary Treatment		
Electricity	$/day	3.66
Operational labor	$/day	78.32
Maintenance labor	$/day	5.96
Repair material	$/day	2.05
Total cost per day	$/day	89.95
Total cost per mg plant influent	$/mg	11.99
Secondary Treatment		
Electricity	$/day	137.00
Operational labor	$/day	98.43
Maintenance labor	$/day	10.05
Repair material	$/day	6.28
Total cost per day	$/day	251.77
Total cost per mg plant influent	$/mg	33.56
Organic Sludge Treatment		
Handling and Dewatering		
Electricity	$/day	16.98
Chemicals—polymer	$/day	175.28
Operational labor	$/day	30.05
Maintenance labor	$/day	26.55
Repair material	$/day	27.57
Total cost per day	$/day	276.45
Total cost per mg plant influent	$/mg	36.86
Total cost per ton dry solids	$/ton	25.40
Incineration and Disposal		
Electricity	$/day	6.04
Natural gas	$/day	99.23
Operational labor	$/day	20.05
Maintenance labor	$/day	10.51
Repair material	$/day	2.68
Total cost per day	$/day	138.52
Total cost per mg plant influent	$/mg	18.46
Total cost per ton dry solids	$/ton	12.71
Total cost Organic Sludge Treatment		
Electricity	$/day	23.02
Natural gas	$/day	99.23
Polymer	$/day	175.28

Table 13-9 (Continued)

	Units	*Costs at 7.5 mgd Flow*
Operational labor	$/day	50.10
Maintenance labor	$/day	37.06
Repair materials	$/day	30.26
Total cost per day	$/day	414.97
Total cost per mg plant influent	$/mg	55.32
Total cost per ton dry solids	$/ton	38.12
Disinfection and Odor Control		
Prechlorination—Odor Control		
Chlorine	$/day	2.18
Maintenance labor	$/day	0.00
Repair material	$/day	0.00
Total cost per day	$/day	2.18
Total cost per mg plant influent	$/mg	0.29
Post Chlorination—Disinfection		
Chlorine	$/day	15.54
Maintenance labor	$/day	1.07
Repair material	$/day	0.07
Total cost per day	$/day	16.69
Total cost per mg tertiary flow	$/mg	1.86
Total Cost Disinfection and Odor Control		
Chlorine	$/day	17.72
Maintenance labor	$/day	1.07
Repair material	$/day	0.07
Total cost per day	$/day	18.87
Total cost per mg treated	$/mg	2.16
Total Operating Cost—Conventional Waste Treatment		
Electricity	$/day	163.65
Natural gas	$/day	99.23
Chemicals—polymer	$/day	175.28
Chlorine	$/day	17.72
Operational labor	$/day	226.85
Maintenance labor	$/day	54.16
Repair materials	$/day	38.66
Total cost per day	$/day	775.57
Total cost per mg plant influent	$/mg	103.05
Operating Cost for Advanced Waste Treatment		
Phosphorus Removal		
Lime Feed		
Electricity	$/day	5.22
Makeup lime	$/day	160.50
Polymer	$/day	25.03
Operational labor	$/day	30.13
Maintenance labor	$/day	8.50

Repair materials	$/day	3.05
Total cost per day	$/day	232.45
Total cost per mg plant influent	$/mg	31.05
Total cost per ton CaO fed	$/ton CaO	*16.37*
Lime Sludge Dewatering		
Electricity	$/day	7.71
Operational labor	$/day	20.05
Maintenance labor	$/day	16.30
Repair materials	$/day	4.74
Total cost per day	$/day	48.80
Total cost per mg plant influent	$/mg	6.50
Total cost/ton CaO recalcined	$/ton CaO	4.42
Lime Sludge Recalcining		
Electricity	$/day	6.66
Natural gas	$/day	145.25
Operational labor	$/day	60.10
Maintenance labor	$/day	15.29
Repair materials	$/day	8.11
Total cost per day	$/day	235.43
Total cost per mg plant influent	$/mg	31.39
Total cost/ton CaO recalcined	$/ton CaO	21.35
Total Cost Phosphorus Removal		
Electricity	$/day	19.60
Natural gas	$/day	145.25
Makeup lime	$/day	25.03
Operational labor	$/day	110.29
Maintenance labor	$/day	40.09
Repair materials	$/day	15.91
Total cost per day	$/day	516.70
Total cost per mg plant influent	$/mg	68.95
Total cost per ton CaO handled	$/ton CaO	45.32
Nitrogen Removal (Based on continuous operation May 1969 through 1969 only)		
Electricity	$/day	61.00
Operational labor	$/day	5.08
Maintenance labor	$/day	5.50
Repair material	$/day	1.38
Total cost per day	$/day	72.96
Total cost per mg plant influent	$/mg	9.73
Recarbonation		
Electricity	$/day	8.49
Operational labor	$/day	16.32
Maintenance labor	$/day	5.18
Repair material	$/day	2.14
Total cost per day	$/day	32.14
Total cost per mg plant influent	$/mg	4.28
Total Cost Nutrient Removal—Phosphorus and Nitrogen		
Electricity	$/day	89.11
Natural gas	$/day	145.25
Makeup lime	$/day	160.50

Table 13-9 (Continued)

	Units	Costs at 7.5 mgd Flow
Polymer	$/day	25.03
Operational labor	$/day	131.70
Maintenance labor	$/day	50.78
Repair materials	$/day	19.44
Total cost per day	$/day	621.81
Total cost per mg plant influent	$/mg	82.97
Filtration (Separation Beds)		
Electricity	$/day	98.60
Chemicals—alum/polymer	$/day	69.09
Operational labor	$/day	23.84
Maintenance labor	$/day	14.28
Repair material	$/day	4.61
Total cost per day	$/day	210.45
Total cost per mg plant influent	$/mg	28.06
Total cost per mg sep. bed effluent	$/mg.	23.57
Carbon Adsorption		
Operational labor	$/day	23.84
Maintenance labor	$/day	3.47
Repair material	$/day	1.52
Total cost per day	$/day	28.84
Total cost per mg plant influent	$/mg	3.84
Total cost per mg tertiary	$/mg	3.23
Carbon Regeneration		
Electricity	$/ton	1.63
Natural gas	$/ton	3.91
Make-up carbon	$/ton	51.79
Operational labor	$/ton	59.24
Maintenance labor	$/ton	40.80
Repair material	$/ton	1.17
Total cost per ton of carbon between regenerations	$/ton	158.70
Total cost per mg plant influent	$/mg	26.86
Total cost per mg tertiary flow	$/mg	15.98
Total Cost for Carbon Adsorption and Regeneration		
Electricity	$/day	2.05
Natural gas	$/day	4.89
Makeup carbon	$/day	64.86
Operational labor	$/day	98.44
Maintenance labor	$/day	18.93
Repair material	$/day	1.96
Total cost per day	$/day	191.16
Mg plant influent	$/mg	25.48
Mg tertiary flow	$/mg	21.41

Total Operating Costs— Advanced Waste Treatment		
Electricity	$/day	189.77
Natural gas	$/day	150.15
Chemicals—alum/polymer	$/day	94.12
Makeup lime	$/day	160.50
Makeup carbon	$/day	64.86
Operational labor	$/day	253.99
Maintenance labor	$/day	84.00
Repair materials	$/day	26.02
Total cost per day	$/day	1,023.43
Total cost per mg plant influent	$/mg	136.51
Total Operating Costs for Conventional and Advanced Waste Treatment		
Electricity	$/day	353.42
Natural gas	$/day	249.38
Chemicals—alum/polymer	$/day	269.40
Chlorine	$/day	17.72
Makeup lime	$/day	160.50
Makeup carbon	$/day	64.86
Operational labor	$/day	400.85
Maintenance labor	$/day	138.16
Repair materials	$/day	64.86
Total cost per day	$/day	1,799.00
Total cost per mg plant influent	$/mg	239.57

Table 13-10 Summary of Total Costs Experienced at South Lake Tahoe.

Process	*Total Operating Costs*	*Capital Costs*	*Total Costs ($/mg influent)*
Primary	12	19.50	31.50
Activated sludge	34	33.60	67.60
Organic sludge	55	14.10	69.10
Chlorination	2	0.30	2.30
Lime treatment	31	9.70	40.70
Lime recalcining	40	13.50	53.50
Ammonia stripping	10	8.00	18.00
Recarbonation	4	4.00	8.00
Filtration	28	17.80	45.80
Carbon adsorption	3	16.30	19.30
Carbon regeneration	27	5.20	32.20
Total of Conventional Treatment	103	67.50	170.50
Total of Advanced Wastewater Treatment	143	74.50	217.50
Total	246	142.00	388.00

All of the costs for actual plant construction and operation were much the same as those projected from pilot plant tests in preliminary estimates prior to design of the full-scale plant. Original estimates are escalated in accordance with the inflation in prices which took place in the interim—that is, between the time the estimates were made and the actual costs were obtained.

ANTICIPATED PERFORMANCE OF VARIOUS PROCESS COMBINATIONS

Due to variations in plant influent municipal wastewater composition from city to city, it is impossible to present predictions of effluent quality and costs of various unit process combinations which would be universally applicable. However, a general indication of expected trends and relative performance is presented in Table 13-11.

The costs of each unit process may be estimated from the preceding section presenting operating and capital costs from South Tahoe for those cases where biological treatment precedes the advanced wastewater treatment plant. In those cases where biological treatment is eliminated and the raw wastewater is treated solely by physical-chemical means, the costs of the activated carbon treatment and the cost of sludge conditioning may be significantly affected, while the costs of coagulation, sedimentation, ammonia stripping, recarbonation, and filtration should not be significantly affected.

It is difficult to project activated carbon costs for physical-chemical processes because of the lack of long-term operating data. Pilot tests indicate that a dosage of 600–800 lb/mg may be typical. The costs presented for the South Tahoe plant were based on a carbon dosage of about 250 lb/mg passed through the carbon. Based on an operating cost of 5 cents/lb of regenerated carbon, the cost of the activated carbon portion of a physical chemical plant would increase by about $25–30/mg over the $50/mg experienced with biological treatment if the above carbon dosage assumption is valid. Of course, this increase would be more than offset by the elimination of the cost of an activated sludge system *providing* the cost of sludge dewatering is comparable. The cost of sludge dewatering is the area of greatest unknown and conflicting data at the present time for physical-chemical systems. Of course, the very substantial space saving provided by the elimination of biological systems is another significant factor to consider in such a comparison.

An examination of Table 13-11 will show that biological pretreat-

Table 13-11 Anticipated Performance of Various Unit Process Combinations.

		ESTIMATED AWT PROCESS EFFLUENT QUALITY						
AWT pretreatment	*AWT process*[2]	*BOD (mg/l)*	*COD (mg/l)*	*Turb. (JU)*	*PO_4 (mg/l)*	*S.S. (mg/l)*	*Color (units)*	*NH_3-N (mg/l)*
Preliminary[1]	*C,S*	50–100	80–180	5–20	2–4	10–30	30–60	20–30
	C,S,F	30–70	50–150	1–2	0.5–2	2–4	30–60	20–30
	C,S,F,AC	5–10	25–45	1–2	0.5–2	2–4	5–20	20–30
	C,S,NS,F,AC	5–10	25–45	1–2	0.5–2	2–4	5–20	1–10
Primary	*C,S*	50–100	80–180	5–15	2–4	10–25	30–60	20–30
	C,S,F	30–70	50–150	1–2	0.5–2	2–4	30–60	20–30
	C,S,F,AC	5–10	25–45	1–2	0.5–2	2–4	5–20	20–30
	C,S,NS,F,AC	5–10	25–45	1–2	0.5–2	2–4	5–20	1–10
High rate	*F*	10–20	35–60	6–15	20–30	10–20	30–45	20–30
Trickling	*C,S*	10–15	35–55	2–9	1–3	4–12	25–40	20–30
Filter	*C,S,F*	7–12	30–50	0.1–1	0.1–1	0–1	25–40	20–30
	C,S,F,AC	1–2	10–25	0.1–1	0.1–1	0–1	0–15	20–30
	C,S,NS,F,AC	1–2	10–25	0.1–1	0.1–1	0–1	0–15	1–10
Conventional	*F*	3–7	30–50	2–8	20–30	3–12	25–50	20–30
Activated	*C,S*	3–7	30–50	2–7	1–3	3–10	20–40	20–30
Sludge	*C,S,F*	1–2	25–45	0.1–1	0.1–1	0–1	20–40	20–30
	C,S,F,AC	0–1	5–15	0.1–1	0.1–1	0–1	0–15	20–30
	C,S,NS,F,AC	0–1	5–15	0.1–1	0.1–1	0–1	0–15	1–10

[1]Prelminary treatment—grit removal, screen chamber, Parshall flume, overflow.
[2]*C,S*—coagulation and sedimentation; *F*—mixed-media filtration; *AC*—activated carbon adsorption; *NS*—ammonia stripping. Lower effluent NH_3 value at 18°C; upper value at 13°C.

ment results in the ability to produce higher-quality effluents by advanced wastewater treatment than by purely physical-chemical means. Several investigators have noted the inability to adsorb on carbon some organics which are removed biologically. Although the physical-chemical approaches now available will produce an effluent quality adequate for severe pollution control situations, the effluent quality required for certain water reuse situations may require a higher degree of treatment provided by biological pretreatment or by new nonbiological techniques.

Using the data in Table 13-11, the cost data for unit processes presented in the previous section, and estimates on sludge dewatering and carbon adsorption costs appropriate to the particular situation, preliminary comparison of alternate treatment plans to achieve a given effluent quality can be made. Adjustment for plant capacity will also be needed. Smith and McMichael (1969) have presented curves describing the relation between plant size and costs for various unit processes. The following table will permit an approximation of costs for various capacities based upon the Tahoe costs presented:

Plant Capacity (mgd)	*Factor to Adjust 7.5 mgd Costs (costs/mg)*
2.5	1.3–1.5
7.5	1.00
50.0	0.7–0.85
100.0	0.6–0.8

Because the effects of plant size vary considerably with each unit process, the above factors should be used only for very preliminary estimates.

References

1. Culp, G. L., and Culp, R. L., "Reclamation of Waste Water at Lake Tahoe," *Public Works* (Feb., 1966).
2. Culp, G. L., and Hansen, S., "How to Clean Wastewater for Reuse," *American City*, June, 1967, p. 96.
3. Culp, G. L., and Slechta, A. F., "Plant Scale Regeneration of Granular Activated Carbon," Final Progress Report, USPHS Demonstration Grant 84–01 (Feb., 1966).
4. ———, "Recovery and Reuse of Coagulant From Treated Sewage," Final Progress Report, USPHS Demonstration Grant 85–01 (Feb., 1966).

5. ———, "Nitrogen Removal From Sewage." Final Progress Report, USPHS Demonstration Grant 86–01 (Feb., 1966).
6. ———, "Plant Scale Reactivation and Reuse of Carbon in Waste Water Reclamation," *Water and Sewage Works,* Nov., 1966, p. 425.
7. Culp, R. L., "Wastewater Reclamation at South Tahoe Public Utility District," *Journal American Water Works Association,* Jan., 1968, p. 84.
8. ———, "Water Reclamation at South Tahoe," *Water & Wastes Engineering,* Apr., 1969, p. 36.
9. ———, "The Status of Phosphorus Removal," *Public Works,* Oct., 1969.
10. ——— and Moyer,, H. E., "Wastewater Reclamation and Export at South Tahoe," *Civil Engineering,* 39 (June, 1969), p. 38.
11. Culp, R. L., and Roderick, R. E., "The Lake Tahoe Water Reclamation Plant," *Journal Water Pollution Control Federation,* 1966, p. 147.
12. Culp, R. L., and Suhr, L. G., "Operations Manual for 7.5 MGD Water Reclamation Plant," Cornell, Howland, Hayes & Merryfield, and Clair A. Hill and Assoc., Sept., 1967.
13. Moyer, Harlan E., "The South Tahoe Water Reclamation Project," *Public Works,* Dec., 1968, p. 7.
14. Nesbitt, J. B., "Phosphorus Removal—the State of the Art," *Journal Water Pollution Control Federation,* p. 791.
15. Slechta, A. F. and Culp, G. L., "Water Reclamation Studies at the South Tahoe Public Utility District," *Journal Water Pollution Control Federation,* 1967, p. 787.
16. Smith, C. E., and Chapman, R. L., "Recovery of Coagulant, Nitrogen Removal, and Carbon Regeneration in Waste Water Reclamation," Final Report of Project Operations, FWPCA Grant WPD–85 (June, 1967).
17. Smith, R., and McMichael, W. F., "Cost and Performance Estimates for Tertiary Wastewater Treating Processes," R. A. Taft Water Research Center Report No. TWRC–9 (1969).
18. Stander, G. J., and Van Vuuren, L. R. J., "The Reclamation of Potable Water from Wastewater," *Journal Water Pollution Control Federation,* 1969, p. 355.
19. Suhr, G. L., and Culp, R. L., "Nitrogen and Phosphorus Removal by High *p*H Lime Coagulation." Presented at the Annual Convention of the Pacific Northwest Water Pollution Control Association, Portland, Oregon (Oct., 1966).

Index